ACTIVE
MATHEMATICS
TEACHING

Research on Teaching Monograph Series

PUBLISHED

Student Characteristics and Teaching by Jere E. Brophy and Carolyn M. Evertson

Pygmalion Grows Up: Studies in the Expectation Communication Process by Harris M. Cooper and Thomas L. Good

Class Size and Instruction: A Field Study by Leonard S. Cahen, Nikola Filby, Gail McCutcheon, and Diane Kyle

Active Mathematics Instruction by Thomas L. Good and Douglas Grouws

The Invisible Culture by Susan Urmston Philips

Cooperative Learning by Robert E. Slavin

FORTHCOMING IN 1983

Staff Networking in Secondary Schools (tentative title) by Philip Cusick

´ACTIVE MATHEMATICS TEACHING,

Thomas L. Good
Douglas A. Grouws
Howard Ebmeier

THE CENTER FOR RESEARCH
IN SOCIAL BEHAVIOR

Longman
New York & London

Active Mathematics Teaching

Longman Inc., 1560 Broadway, New York, N.Y. 10036
Associated companies, branches, and representatives
throughout the world.

Copyright © 1983 by Longman Inc.

Developmental Editor: Lane Akers
Editorial and Design Supervisor: Diane Perlmuth
Manufacturing Supervisor: Marion Hess
Production Supervisor: Ferne Y. Kawahara

Library of Congress Cataloging in Publication Data

Good, Thomas L., 1943–
  Active mathematics teaching.
  (research on teaching monograph series)
  Includes index.
  1. Mathematics—Study and teaching (Elementary)—
Missouri.   2. Mathematics—Study and teaching
(Secondary)—Missouri.   I. Grouws, Douglas A.
II. Ebmeier, Howard.   III. Title.   IV. Series.
QA135.5.G59   1983   510'.7'10778   82–4646
ISBN 0–582–28342–6

Manufactured in the United States of America
Printing: 9 8 7 6 5 4 3 2 1   Year: 91 90 89 88 87 86 85 84 83

Contents

v

Preface

We have attempted to write this book for a general audience because we feel that the book has considerable value for those interested in education generally as well as for practitioners and classroom researchers. In the text, we refer readers to published materials where they can obtain more technical information; for readers who are interested in more information about training and observational procedures, all three final reports prepared for the National Institute of Education are available in the ERIC system.

Developing the Missouri Mathematics Program has been a part of our lives for several years. Although we have been concerned with many other research programs and a variety of teaching, writing, service, and administrative duties, we have thought about and worked on this program for some time. Indeed, the first two authors' involvement in this research began about a decade ago and the third author has participated in the research program for over five years.

Given the span of time and the number of projects that have been completed in our program, it is impossible to identify and thank all the people who have assisted us. However, we do want to express our gratitude to a few key individuals and organizations who contributed directly and indirectly to the publication of the book. Significant support was provided from the National Institute of Education (NIE) in the form of three large research grants. Without this financial support the research could not have been completed. Also, the personal support and encouragement from several staff members at NIE was appreciated. Although we express our thanks to NIE for financial support, it should be clearly understood that the ideas we present here are our own and do not necessarily reflect the opinion or the endorsement of NIE.

The Center for Research in Social Behavior, the graduate school, and the college of education at the University of Missouri-Columbia have given considerable support to our research activities. Several administrators have provided both direct and indirect support for our efforts. We want to thank in particular Bruce J. Biddle at the Center for Research in Social Behavior, Don Blount and Richard Wallace in the graduate school, and Bob Woods and Wayne Dumas in the college of education. The graduate school has been and continues to be a strong force of support for social science research, and we have benefited from that support in numerous ways. We have in particular benefited from the presence of the Center for Research in Social Behavior, a research unit sponsored by the graduate school. The center and its director, Bruce J. Biddle, have provided invaluable assistance through the provision of office space and countless personnel services.

Countless drafts of technical reports, letters, articles, final reports, observational manuals, chapters (and now a book) related to the project have been typed by several secretaries at the center. Terry Brown, Sherry Kilgore, Janice Meiburger, and Patricia Shanks have transformed an unending stream of dictation tapes and yellow pads (filled with virtually illegible handwriting) into polished manuscript. Their work has been of the highest quality and their contribution to the research program has been major and invaluable. Most of the work on the book manuscript per se has been done by Janice Meiburger and Particia Shanks and we especially thank them for their professional efforts.

Gail Hinkel has provided immense support to the project as an editor. She has suffered through countless rough drafts and redrafts of articles, chapters, and reports. She has provided consistent help in enabling us to communicate our findings and ideas more clearly. She has lent considerable expertise and knowledge to our efforts, and we are extremely grateful to her for the many professional contributions she has made.

Billye Adams and Barb Breen helped us with a variety of tasks (payroll records, ordering supplies, arranging for forms to be printed, etc.). We thank them for their support and help.

We want to express our genuine gratitude to a number of individuals who helped us in various stages of conducting, analyzing, and preparing project materials: Terrill Beckerman, Sally Beckerman, John Engelhardt, Larry Flatt, Carol Flatt, Ruthanne Harre, Gail Hinkel, and Sharon Schneeberger. Terrill Beckerman played an especially important role in supervising observational training and data collection as well as much of the data analysis.

Finally, we wish to thank our families who gave us the extra time called for in writing this book. Their interest and respect for our research task was a source of considerable support.

* * *

Acknowledgment is also due to the following sources for permission to reprint:

Permission has been granted to reprint from Short, Edmund C., "Knowledge Production and Utilization in Curriculum: A Special Case of the General Phenomenon," *Review of Educational Research*, Summer 1979, pp. 237–301, copyright 1979, American Educational Research Association, Washington, D.C.

Permission to reprint from Terrill Beckerman's 1981 Ph.D. dissertation at the University of Missouri, "Effects of Student Type and Math Programs Upon Performance," has been granted by the author.

Permission to reprint from Nichols, Anderson, Dwight, et al.: *Holt School Mathematics, Grade 4* copyright © 1974 by Holt, Rinehart and Winston, Publishers, has been granted by the publisher.

Permission to reprint from Dr. Ruthanne Harre's 1980 Ph.D. disseration entitled "Interaction Effects on Time on Task in Mathematics" has been granted by the author.

Permission to reprint table 4.2, 4.3, 4.4 from *The Journal of Educational Psychology*, 1979, 71, 355–562, copyright 1979 by The American Psychological Association has been granted by the publisher.

Permission to reprint from *Handbook of Teaching and Policy* edited by Lee Shulman (forthcoming from Longman Inc.), has been granted by the publisher.

1

Introduction/Overview

The purpose of this book is to describe the Missouri Mathematics Effectiveness Project, a major systematic research program that was conducted in the mid-1970s and early 1980s and supported by the National Institute of Education. The research reported is based upon a decade of naturalistic and experimental study of mathematics classrooms in intermediate elementary grades and in junior high classrooms. We will present a rationale for the particular research questions we chose to address, describe the research procedures and findings from several studies, and discuss the implications of our research for teachers, teacher educators, and classroom researchers.

When we began our program of research in the early 1970s, there was very little useful or reliable information available for describing the relationship between classroom processes (e.g., teacher behavior) and classroom products (e.g., student achievement). What knowledge existed in 1970 about the effects of classroom processes on student achievement was weak and contradictory. After a decade of extensive research on classroom processes (much of this research supported by the National Institute of Education), there is now much pertinent information about this relationship.

Similarly, in the past ten years the literature on basic skill instruction in reading and mathematics in elementary schools has moved from a state of confusion to a point at which experimental studies can be designed upon a data base. Several large-scale correlational studies produced data which illustrated that it was possible to identify some teachers who consistently obtained more student achievement than expected (e.g., Brophy and Evertson, 1974; Good and Grouws, 1975a). Furthermore, research showed that it was possible to identify instructional patterns that differentiated these teachers from teachers who were not as successful, according to a given operational definition of effectiveness (e.g., Berliner

1

and Tikunoff, 1976; Brophy and Evertson, 1976; Rosenshine, 1979). Finally, certain patterns of effective instructional behavior suggested by the correlational studies have been supported by field-based experimental studies (e.g., Anderson, Evertson, and Brophy, 1979; Good and Grouws, 1979; Stallings, 1980).

In this chapter we want to suggest why recent research has been more successful than previous attempts to associate teaching acts with student achievement. Despite the considerable progress that has been achieved, much more information is needed about how (and why) instructional behaviors influence student learning. Later in the book, we will discuss some forms of research which are needed in the future (especially mathematics research).

In this chapter we will discuss a belief that characterized the literature in the late 1960s and early 1970s, the belief that teaching does not (or cannot) make a difference in student learning. We also want to briefly summarize some of the issues we addressed that were related to earlier studies of teaching. Certainly the problems associated with previous research influenced the way in which we defined our objectives and conducted our research program. Finally, in this chapter we will provide an overview of the studies that will be presented in the remainder of the book.

Do Teachers Make a Difference?

Over 20 years ago, Gage (1960) pointed out that more than 10,000 studies had been conducted on the topic of teacher effectiveness, and noted that the combined literature was overwhelming, both in terms of the number of studies and the inconsistencies across findings. Others, too, have commented upon the size of the literature and its limited scientific and practical value. As a case in point, the Committee on Teacher Effectiveness of the American Educational Research Association presented the following conclusion about 30 years ago:

> The simple fact of the matter is that, after forty years of research on teacher effectiveness during which a vast number of studies have been carried out, one can point to few outcomes that a superintendent of schools can safely employ in hiring a teacher or granting him tenure, that an agency can employ in certifying teachers, or that a teacher-education faculty can employ in planning or improving teacher-education programs (Committee, 1953, p. 657).

Dunkin and Biddle (1974) reviewed this literature and reached a similar conclusion about two decades later: there was little reliable information about the relationship between teaching behavior and student achievement.

Why is it that the results from research on teaching were inconsistent and contradictory until the mid 1970s? In the past, a frequent explanation was that teachers are not differentially effective. That is, students learn and integrate new material because of variables that largely have nothing to do with teachers or classroom environments (e.g., student aptitude, degree of parental support, and interest in learning). For a detailed statement of this view, see Stephens (1967).

Other social scientists have argued that the poor design of most early educational research almost guaranteed that relationships between teacher behaviors and student achievement could not be found. Dunkin and Biddle (1974) contended that a basic problem was that teachers were seldom observed. If one is to argue, as Stephens did, that teaching behavior does not make a difference in student learning, it is important to present information (1) that all teachers teach the same way, or, (2) that differential teaching behavior has the same influence on student outcomes.

Unfortunately, when we began this program of research in the early 1970s, there were few studies that included data on classroom observation (e.g., teacher behavior) and student outcomes (e.g., achievement), and it was impossible to reach firm conclusions about whether teachers made a difference in student learning. Even large and reasonably sophisticated studies were characteristically flawed by a failure to include observation of classroom and/or school processes. For example, the massive (involving over 4,000 public schools and 645,000 students) report by Coleman et al. (1966) lacked observational data. The study presented data which have been interpreted by some as proof that schools have little effect upon student learning and that additional expenditures are unlikely to improve student performance.

Like most previous teacher effectiveness research, the Coleman report can be criticized because it deals with input and output variables but not with classroom processes. A related criticism is that this research considered pupil achievement in relation to school differences rather than teacher differences. If teachers make a difference, and if pupils are to be exposed to both good and poor teaching, then it should not be surprising that many schools are found to be quite similar in their effects on students (Good, Biddle, and Brophy, 1975).

Need for New Research

Despite the paucity of data describing what took place in classrooms (e.g., Schwab, 1969), it was almost becoming a national sport to criticize teachers and schooling in the late 1960s and early 1970s. Popular books like *Death at an Early Age* (Kozol, 1967) or *Crisis in the Classroom* (Silberman, 1970), among many others, were frequently cited as evidence that schools or teachers made little difference in the educational advancement of students. Unlike much previous classroom research and writing, these books did attempt to comment upon classroom process. However, such books were based upon practices of one or a few schools and therefore could not accurately describe what was occurring generally in American schools. When school critics did attempt to sample schools more broadly, they seldom documented the observational procedures used to collect information and the reliability and validity of their conclusions were unknown.

Implicit in much of the school critics' writing was the argument that there appeared to be little variability in school and/or classroom routines. To see one school or classroom was to see them all. We doubted the veracity of this argument for two reasons. First, in our own classroom observations we had found immense variability in the behavior of classroom teachers and we felt that some forms of teacher behavior were more effective than others. We therefore became interested in testing the hypothesis that teachers can make a measurable difference in students' learning. Second, we thought that critics' advocacy of immediate and sweeping solutions to the problems of American schools was doomed to failure because solutions were not likely to be based upon knowledge of what actually took place in American classrooms. Hence, we felt that the potential effects of good teaching on student performance were consistently underestimated by school critics and by recent research and policy evaluations (e.g., Coleman et al.; the early Head Start evaluations). In particular, it was important to determine whether natural variation in schools (e.g., differences in teacher behavior) could be associated with student achievement before making substantial changes in schooling.

Our initial goal, then, was to collect data relevant to the charge that there was limited variability in teacher behavior and that teachers had little effect on student achievement. We ultimately became interested in altering and trying to improve certain aspects of classroom practice. Clearly, classroom research does not occur in isolation from society, and our reasons for studying the "teacher difference" hypothesis were related to the social context of the early 1970s. We did not feel that critics were totally wrong about the state of American schooling. We knew there was (and is) much that could be improved in American classrooms. However, we thought that it was important to challenge the idea that teaching made little difference in students' learning.

Improving upon Previous Research

As noted above, one of the reasons that previous classroom research was unreliable was that few observational data were collected to describe teaching. Many investigators assumed that they knew what effective teaching was and attempted only to verify their definitions. Previous researchers had measured only a few very general aspects of teaching. This explained why consistent relationships between teaching behavior and student outcomes had not been uncovered and why the general "sameness" of teaching had been overestimated (e.g., teachers talk a lot). One important way to improve upon previous teaching effectiveness research was for us to assess classroom behavior as comprehensively as possible. (Our procedures for doing this are provided in Chapter 2.)

However, since we thought that no two classrooms or teachers were the same, it was important that we choose our observation sample with considerable care. Unfortunately, most previous observational samples consisted of volunteer teachers who were available and convenient. Thus, even in the few studies that included relatively extensive observation of teachers, there was no reason to assume that any of the teachers were effective at obtaining identifiable student outcomes. Mitzel and Gross (1958) reviewed several studies that used pupil growth criteria as an index of teacher effectiveness and noted that authors of such studies had provided no assurance that "effective" teachers were effective in any absolute sense. At best, teachers' effectiveness was argued on the grounds that they were effective in comparison to other teachers in the sample (and often a very small sample). Furthermore, even the criterion of relative effectiveness was often weakened by the tendency in such research to study first-year teachers. Understandably, such studies produce conflicting, weak findings about the relationship between selected teacher behaviors and student outcomes. In our research, we wanted to incorporate variation in teachers' ability to produce scores on standardized achievement tests. That is, we attempted to assure ourselves that teachers did differ in their effects upon students before we entered the classroom to observe them (details are presented in Chapter 2).

Another problem with past research on teaching was its focus on generalized aspects of teaching and the related failure to consider the effects of a particular context on the assessment of teaching effects (e.g., Dunkin and Biddle, 1974). In addition to observing a group of teachers who were carefully selected before observation, we wanted to restrict our sample to grades three and four, and to mathematics. Although a study that focused all classroom observation on a specific context was an improvement over previous research, we knew that resulting statements about effective teaching might be limited to the context studied.

To reiterate, the difficult goal of identifying patterns of teacher behavior which affect student learning made it desirable to focus observation upon classroom activity during the teaching of a particular subject. We chose mathematics because of its importance in the elementary school curriculum (reading and mathematics are commonly accepted as the major curriculum areas in elementary schools). We were also better able to study math technically (one of the authors is a specialist in mathematics education) and we felt that more teacher and school variance would be associated with students' mathematics performance than reading performance. This assumption has now received empirical support from Coleman's analyses of data from the International Educational Study (1975).

Teacher Effectiveness

Earlier research had also suffered because criteria for effectiveness were vague or unspecified. To compare teachers in terms of their effects on students' mathematics achievement, it was necessary to develop an operational definition of effectiveness. We defined teacher effectiveness as student performance (residual gain) on a standardized achievement test. Although this is not a complete definition it is *one* important aspect of effective teaching, and there is a reasonable consensus among teachers and standardized test developers about what constitutes the basic mathematics curriculum in elementary schools and in beginning junior high mathematics classes. Standardized achievement scores can be a partial criterion of effectiveness if one understands their limitations and does not overgeneralize findings based upon them.

Still, teaching is composed of many varied skills, and teachers have numerous duties to perform in the classroom. Although students' achievement and understanding of mathematics concepts are major goals, they are not the exclusive tasks of teaching. Teachers vary in "effectiveness" on different tasks and, depending upon the definition of teacher effectiveness a researcher employs (e.g., classroom climate), a different rank order of "effective" teachers will emerge. The definitional problem of effectiveness is present in any comparison. For example, if someone raised the question, Is he/she a good basketball player, we would have to have some definition of "goodness" to answer the question. That is, are we referring to the ability to throw a ball, dribble a ball, shoot a ball, or one of several other criteria that could be invoked?

Another possible limitation of a standardized test is that it may not match closely the content presented in students' textbooks or the curriculum that teachers cover in their classrooms. In our research in both elementary and secondary classrooms we ascertained that the standardized test adequately reflected the curriculum and that the test was a reasonable measure of student learning in both treatment and control classrooms (details will be presented later in appropriate chapters).

In the chapters that follow, we will often refer to effective and ineffective teachers; however, we do so only in a relative and restricted sense. Specifically, we mean that effective teachers are getting higher mean student standardized achievement gains than other teachers who teach comparable students under similar circumstances. Had we used different operational procedures, even on the same measure, we might have identified a different set of teachers as being effective. For example, instead of looking at mean achievement gains, it would have been possible to study increases or decreases in students' achievement variation within a classroom (before and after the instructional year). That is, are students closer together or farther apart in achievement at the end of a year with one teacher than a comparable group of students with a different teacher?

As will be seen in Chapter 2, we originally tried to identify a sample of teachers who had very positive and stable effects on students' achievement and attitudes over consecutive years. However, we did not find enough teachers who had extreme and stable effects on both outcomes. In the original study we therefore observed only teachers who were consistently able to get higher achievement gains than other teachers instructing comparable students.

However, in the original study and in subsequent research we have measured students' and teachers' attitudes toward mathematics (as well as toward the specific mathematics program we developed later), and these affective reactions are included in our interpretation of the data. Although initial emphasis was on a teacher's effect on classroom mean achievement (How much did the students as a group gain during the year?), as the research progressed we raised important questions about the effects of particular types of teaching on different types of students.

OVERVIEW OF THE RESEARCH

Chapters 2 and 4 demonstrate that our efforts to systematically observe a carefully selected group of teachers in a highly specified context did answer our initial question: "Can variance in teacher behavior be related to student achievement?" However, the question, "Do teachers make a difference?" led us to examine a number of aspects of mathematics teaching that we had not contemplated when we designed our first study. Before describing the details of procedures and methodological issues it will be useful to present in very general terms the entire program of research.

Naturalistic Study

The purpose of the original study was to determine whether it was possible to identify teachers who were consistent (across different groups of students) and relatively effective or ineffective, using student performance on the Iowa Test of Basic Skills as an operational criterion. We wanted to observe teachers who differed in effectiveness and to see if differences in their classroom behavior could be identified.

In brief, high residual mean scores appeared to be strongly associated with the following teacher behaviors: (1) large-group instruction; (2) generally clear instruction and availability of information to students as needed (process feedback, in particular); (3) a nonevaluative and relaxed learning environment which was task focused; (4) higher achievement expectations (more homework, faster pace); and (5) classrooms which were relatively free of major behavioral disorders. Teachers who obtained high student achievement test scores were active teachers. They gave a meaningful and clear presentation of what was to be learned, provided developmental feedback when it was needed, structured a common seatwork assignment, and responded to individual students' needs for help.

Preparing for Experimental Study I

We were pleased that some consistent differences could be found in correlational research between relatively effective and ineffective mathematics teachers. However, at that point we only had a description of how more and less effective teachers (in our sample) behaved differently. We did not know if teachers who did not teach the way more effective teachers did could change their behavior or whether students would benefit if teachers were trained to use new methods. To answer these questions, we developed a training program (describing how effective teachers behaved in the naturalistic study) and conducted an experimental study to determine what effects the program would have on teacher behavior and student achievement.

In writing the teacher training materials and instructional program, our earlier naturalistic findings were integrated with the recent naturalistic research of others and with existing experimental research in mathematics education. Some of the variables we tested did not come directly from teaching behaviors measured in our earlier studies, but were instead based upon what observers had seen in classrooms. Still other variables (e.g., mental computations) were based on findings from experimental studies. The program is a *system of instruction:* (1) instructional activity is initiated and reviewed in the context of meaning; (2) students are prepared for each lesson stage to enhance involvement and to minimize errors; (3) the principles of distributed and successful practice are built into the program; (4) active teaching is demanded, especially in the developmental portion of the lesson (when the teacher explains a concept being studied, its importance, etc.).

Experimental Study I

In the naturalistic study, emphasis was placed upon internal consistency. We chose a relatively stable school district in order to exclude as many rival hypotheses as possible to the conclusion that teachers and teaching were affecting student learning. In the initial experimental study, a more heterogeneous school population was sampled because we felt this would be a more stringent test of the training program.

Observers' records indicated that the experimental teachers implemented the program very well (with the exception of certain recommendations concerning how to conduct the developmental portion of the lesson). Pre- and posttesting with the SRA standardized achievement test indicated that after two and one-half months of the program, students in experimental classrooms scored five months higher than those in control classrooms. Results of a content test which attempted to more closely match the material that teachers were presenting than did the standardized tests also showed an advantage for experimental classes.

Pre- and posttesting on a ten-item attitude scale revealed that experimental students reported significantly more favorable attitudes at the end of the experiment than did control students. Also, it is important to note that anonymous feedback from teachers in the project indicated that they felt the program was practical and that they planned to continue using it in the future. Research elsewhere indicates that teachers have a favorable reaction to the program, even when it is presented and discussed without the involvement of the developers (Keziah, 1980; Andros and Freeman, 1981).

To explore achievement patterns more fully in terms of student and teacher characteristics, it was considered important to define teacher and student types more broadly. To develop student typologies, an instrument (aptitude inventory) was designed to assess student characteristics which might interact with key features of the treatment program, identifiable teacher characteristics, and/or classroom procedures. To obtain teachers' views of the characteristics, organization, and typical activities of their classrooms, a questionnaire was developed (teaching style inventory). The aptitude inventory was administered to all students in the sample and the teaching style inventory was administered to each teacher.

Results suggested that the treatment generally worked (i.e., the means in each cell were in favor of the treatment group), but the program was more beneficial for certain combinations of teachers and students than for others. The data collectively indicated that teachers who implemented the model got good results, yet some teacher types used more facets of the program than did other teachers.

One of the most interesting findings of the study was the interactions between teacher type and treatment type. There was a strong teacher effect in the treatment condition that was not found in the control sample.

Although the results of the experimental field study have strong implications, they must be interpreted in light of the evidence that the experimental treatment worked better for some combinations of teachers and students than for others. Still, students generally benefited from being in treatment classrooms.

Experimental Study II: Verbal Problem Solving

Following this successful experiment we debated several possible topics for further study. Ultimately, we decided to study verbal problem solving and to see if teacher behavior and student achievement in this area could be affected. We felt that if mathematics knowledge is to be applied to "everyday life," students need practical problem-solving skills (e.g., the ability to determine whether a 12-ounce or 16-ounce package is the better buy). Unfortunately, extant literature on instructional behavior and students' performance on verbal problem solving did not provide a data base (drawn from naturalistic observations in classrooms) for building a treatment program. Because we thought it was important to understand and to possibly improve students' abilities to solve relatively simple verbal problems, we decided to make a systematic effort to develop testable instructional strategies. We therefore broadened the instructional program by adding a section on verbal problem solving, and by writing a training manual detailing instructional strategies which teachers might use to teach students verbal problem-solving skills.

The second field experiment was conducted one year after the first, in the same school district. The expanded program (the training manual used in experiment one plus the verbal problem-solving manual) was evaluated in 36 sixth-grade classrooms, and the general design and training procedures were the same as those utilized in the first field experiment.

The only exception to the similarity of conditions between field experiments one and two was a major one. In the first field experiment, all teachers were using a semidepartmentalized structure (teachers taught only two or three different subjects a day). In the second field experiment, three organizational patterns were represented in the teacher sample. Some teachers utilized the semidepartmentalized structure; other teachers taught only math as a special subject (sixth-grade teachers taught math to several different sixth-grade classes); and some teachers were in open classes (where team teaching and individualized instruction were prevalent).

The raw means and standard deviations for the SRA (pre- and posttest) and the problem-solving posttest, by treatment condition and by organizational structure, indicated that student performance increased from pre- to post- in all cases on the 40-item SRA test. Furthermore, all treatment groups surpassed the performance of equivalent control groups. Two of the three treatment groups had higher mean performances than equivalent control groups on the problem-solving test. It should be noted that the exception, the open-treatment classes, had the lowest pretest score on the SRA.

The results of the formal analyses using adjusted mean scores indicated that the performance of the treatment group was not significantly higher than that of the control group on the post-SRA test, using the pre-SRA test as a covariant (with all forms of classroom organization included in the analysis). A similar analysis was performed on the problem-solving test (using the pre-SRA test as a covariant) to compare the significance of adjusted means across all treatment and control classrooms. This analysis indicated that the performance of the treatment group exceeded that of the control group in a way that approached significance. Furthermore, when open-space teachers were excluded from the analysis, the comparison on the problem-solving test revealed that the treatment group's performance was significantly superior to that of the control group ($p = .015$).

Reactions of the treatment teachers were assessed confidentially two months after the program ended. The overall reaction of experimental teachers was extremely positive. Questionnaire responses revealed that two-thirds of the participants continued using all aspects of the program at or very near the initial level recommended by the project directors. After the program ended, 10 teachers were still including verbal problem solving in their curricula and 13 were implementing the prescribed development phase at least three times a week. Fifteen teachers continued to assign homework a minimum of three nights a week, and 13 were conducting weekly and monthly review sessions.

At the debriefing session we provided control teachers with a copy of the program manual. Two months later we assessed their reactions to these materials. We did this for two reasons. First, we wanted to see how teachers who had been exposed to the program but did not use it would evaluate it. Were the favorable comments of experimental teachers due to the fact that they had used the program and hence felt obligated to recommend it? We also wanted to see how new the various aspects of the program were to control teachers. Their responses indicated that they were familiar with most of the recommended teaching techniques, and two or three of the control teachers said their supervisors had advocated that they use a directed lesson. Such responses suggest to us that many of the control teachers were using some instructional techniques which were similar to those in our program.

Seventeen of the 19 control teachers responded to the questionnaire. Five control teachers reported they had carefully read the general manual and the verbal problem-solving manual. Five others had read both manuals quickly and six had at least skimmed them quickly and had thought about the highlights. Responses revealed that there was considerable correspondence between the teaching methods control teachers were already using and those requested by the program. Eight teachers reported they were already utilizing the prescribed development and seatwork aspects of the program, and were also teaching their classes as a whole. At least five more teachers reported general overlap between the program and what they had been doing, for each category except the verbal problem solving. It seems reasonable to suggest that the general achievement of students was not enhanced by the experimental program partly because control teachers were using many aspects of the program, although we do *not* have observational data to verify this point. Still, it is clear that the verbal problem-solving material was basically unique to control teachers (the new part of the present program and the part that was disseminated in the school system only to experimental teachers in this study). The experimental program did have positive effects upon students' verbal problem-solving skills.

Experimental Research in Secondary Schools

Considering the relatively successful results of the two experimental studies in elementary schools, we were very much interested in expanding our inquiry to secondary classrooms. There was comparatively little process data which described mathematics teaching in secondary settings. What data did exist at the secondary level were consistent with the research we had conducted and the program we had built at the elementary school level. We had hoped to have the opportunity to sample secondary teachers' reactions to the program directly before modifying it.

Originally we had planned a three-year project in which we would first conduct a treatment study (simply asking mathematics teachers to implement the existing program) and use the results and the responses of these teachers (after they had utilized the program) to build a modified mathematics program for use in secondary settings. A second study would actively involve other teachers in the modification of the training program. That is, another group of secondary teachers would be provided with program materials and all related findings, including the results of the first secondary study, and comments by previous teachers who had used the program in fourth-, sixth-, and eight-grade classrooms.

Unfortunately, the research program had to be modified to fit into a period of 18 months (as opposed to three years) and because of the time limit, it was possible to conduct only one major treatment study. Since we were committed to the idea of involving teachers in reviewing and planning the research, the question was, Under what circumstances could we do this? We found out about reduced funding in the summer and had to address the issue of involving teachers at that time. We had two alternatives: we could work quickly with the teachers and be able to begin the treatment in the fall, or we could use the fall semester to become acquainted and work with teachers, and conduct the experiment in the spring. We chose the former, less involved arrangement. We thought that secondary teachers might be less responsive than elementary school teachers to research (as described in the popular literature), and that once routines (and plans) were established in the school year, both secondary teachers and students would be more hesitant about changing classroom procedures.

We ultimately excluded from consideration models which would provide continuing feedback to teachers (the chance to observe or to be observed by fellow teachers) or give them the opportunity to modify the program in major ways once it had been initiated. After deciding to test a treatment program relatively early in the year, we used a minimal partnership arrangement wherein the time for involvement between teachers and researchers was relatively limited, but the decision-making process was still open and all aspects of the program were subject to change.

Nineteen teachers from 12 different junior high schools volunteered for the project. Most of the target classrooms in the study were regular eighth-grade mathematics classrooms. Of these, six were assigned to the partnership group. After reading the program description, the six partnership teachers met with project investigators and discussed and modified the program (procedures at the meeting and teacher-suggested program changes are discussed in the text). Five teachers were assigned to the treatment group. Both the partnership teachers and the treatment teachers were asked to use the instructional program in their classrooms. The only difference between the two groups was that the partnership group had a chance to modify the program, and the treatment group did not.

Some teachers implemented the program more fully than others. Among many findings were the following: (1) the average implementation score was found to correlate significantly with students' attitudes toward mathematics, and (2) instructional time spent on verbal problem-solving activities did correlate significantly with students' problem-solving achievement scores. Finally, students' performance in verbal problem solving in both partnership and treatment classrooms was superior to problem-solving performance in control classrooms. Students' general computational achievement was not affected by project participation.

In addition to discussing the procedures and findings of the secondary study in detail in this book, we also discuss what we have learned about doing research with teachers and how we would attempt such partnership research in the future.

SUMMARY

Across the several studies conducted at different grade levels, we believe that we have identified many aspects of mathematics teaching that researchers and practitioners need to consider. In the chapters that follow we will argue that teachers make a measurable and important difference in the way in which students learn mathematics. Our research provides compelling evidence that teachers make a difference in student learning and offers some useful information about how more and less effective teachers differ in their behavior and in their effects on student achievement. In particular, we will argue that the concept of active teaching is useful for studying classroom teaching and learning.

2

The Naturalistic Study

BACKGROUND

From 1973 to 1975, with the support of the National Institute of Education, we conducted a large observational study of teaching effectiveness in fourth-grade mathematics classrooms (for full technical details, see Good and Grouws, 1975a). The purpose of this research was to determine whether it was possible to identify teachers who were consistent (across different groups of students) and relatively effective or ineffective, using average student performance (classroom mean) on the Iowa Test of Basic Skills as an operational definition of effectiveness. Furthermore, we intended to observe teachers who differed in effectiveness and to see if we could identify teacher behavioral patterns that were related to increased student achievement.

As we noted in Chapter 1, teacher effectiveness research had been a popular but unproductive activity in the past. Most of this research did not include observation of teaching, or had done so narrowly with pre-conceived notions about how teachers should behave in the classroom. Furthermore, most teacher effectiveness research had employed criteria for assessing teacher effectiveness that lacked validity or generality.

In this chapter we will discuss our initial research study and indicate how we attempted to solve some of the problems of past effectiveness research. Specifically, we will present our procedures for selecting a sample of teachers; observational procedures used to study and to differentiate the behaviors of relatively high and low teachers; and the behavioral findings that characterized the classroom behavior of relatively effective and ineffective teachers.

FOCUS ON MATHEMATICS

Because we believed that behaviors and principles which characterized effective teaching would vary from context to context (teaching reading in first-grade classrooms vs. mathematics teaching in sixth-grade class-rooms), it was desirable, at least· initially, to focus observation upon a particular classroom context. Mathematics was chosen primarily because of its importance in the elementary school curriculum (reading and mathematics are commonly accepted as the core curriculum in elementary schools). Also, we believed that factors associated with teachers and schools would have more of an effect upon students' mathematics performance than their reading performance (e.g., parents often teach children verbal skills but are less likely to teach mathematics skills).

Our initial focus on third- and fourth-grade classrooms was due to our interest in determining if and how teachers affected student learning. We felt that few parents were teaching the mathematics content emphasized in the modern curriculum at grades three and four (i.e., many parents are not familiar with mathematics curriculum emphasizing sets, geometry, etc., and most show little enthusiasm for learning or teaching such content) and thus home background influence would be minimal.

STABLE POPULATION

As we emphasized in Chapter 1, our original and most important objective was to test the hypothesis that teachers make a difference in mathematics instruction. Hence, in selecting our original sample we tried to control some of the obvious factors that might confound findings that certain instructional behaviors were more effective than others.

Unless teachers are teaching students with similar backgrounds (e.g., students in all classrooms being compared are similar in SES), it would not be possible to argue that certain teachers appear to obtain higher levels of student achievement than other teachers. To control for differences in student population, classrooms can be deliberately matched for relevant student characteristics (unfortunately, researchers seldom have the opportunity to affect student assignment to particular classrooms), or researchers can capitalize upon such classroom similarities when they occur naturally.

Fortunately, we were able to obtain a large number of classrooms that contained reasonably comparable students. We had access to student achievement records from a large school district that bordered a metropolitan area in the Midwest. The district served a largely middle-class population, and there was comparatively little SES variance across schools. Thus, through naturalistic and statistical controls (regression procedures), which will be discussed later, we could judge the relative effectiveness of a large number of teachers.

There were other factors that made the sample a good one for our purposes. The stability of teaching conditions appears to be a very important and relevant factor, but one seldom discussed in the teaching effectiveness research literature. If the conditions of teaching are not stable, it is difficult to make inferences about teacher effectiveness. For example, if the student population in schools under study is changing rapidly, then fluctuation in student achievement may result from changes in the student population and not from teacher behavior. We were able to conduct our study in a district that had a very stable student population (the average aptitude of students in the school district had been the same for over a decade).

If the faculty of a school being sampled changes from year to year, it is very difficult to make inferences about the effectiveness of a teacher from that school. For example, if a third-grade teacher receives students from the classrooms of two second-grade teachers one year (who emphasize verbal problem solving) and the following year receives students from two different second-grade teachers (who emphasize computation), then students' scores in the third-grade classroom may reflect their experience in the second-grade classroom as well as what transpires in the third-grade classroom.

If teachers vary in effectiveness from year to year, it may be due to the fact that they have classes with different types of students from one year to the next or because their students have been exposed to diverse types of instruction in previous years. Our sample and general design made it more likely that any differences in student achievement could be attributed to teaching behavior. Although there is no way to control for all the complicated relationships among variables that affect student achievement in a single year, we tried to obtain a sample of teachers from schools which had relatively stable teaching conditions.

We also tried to control for differences among schools in textbooks and curriculum materials. At the time we conducted the initial study, it was not uncommon for school districts to allow individual schools and teachers to choose whatever textbooks they wanted (or at least select one from an approved list of materials) for instructional purposes. However, it is difficult to compare the relative teaching effectiveness of teachers who use different curriculum textbooks. Instructional pace and order of subject matter presentations which characterize specific curriculum material may interact with teaching method. If we found that some teachers were achieving better gains than other teachers and we had not controlled for curriculum material, it would be difficult to determine whether instructional behavior, curriculum materials, or some combinations were more likely to be associated with increased student achievement. It was therefore advantageous to study teachers (as we did) who were using the same curriculum materials, because we wanted to rule out the possibility that the textbook affected students' achievement.

To reiterate, selecting a school district that had a relatively homogeneous student population and reasonably stable teaching conditions would maximize the internal validity of the study and provide a more rigorous and direct test of teacher effects on student performance as measured by standardized achievement tests. Although these strict controls might weaken the external validity of the study (the application of these results to other settings), this was of no major concern to us because we intended to test the external validity of the results in subsequent studies.

Research which defines teacher effectiveness as student gain usually depends on multiple regression or covariance techniques to adjust class gain scores for variability in students' initial performance levels. At first glance, this appears to be a logical method of studying the effects of teachers on student performance. Regression techniques statistically control differences in initial (prior to instruction) student behavior; thus, it is argued that the differential student gains in Classroom A "must be due to Teacher A" because child differences are statistically removed. However, Teacher A in September may have 20 of 30 third graders above grade level; Teacher B may have only 6 of 30 students above grade level. Statistical controls are not directly related to important differences between such classrooms (the incidental learning and motivational climate associated with the grouping of good students in the same classroom, for instance). Regression analysis is the appropriate way to approach this problem; however, such initial differences between classes should be minimized as much as possible through other means. Classrooms in which conditions vary greatly cannot be equated through statistical means.

There were only a few earlier studies which had employed methodology to assure that teachers being studied were teaching under quasi-similar conditions (Torrance and Parent, 1966; Soar, 1966). At the time we began our research several other investigators (e.g., Jere Brophy and Carolyn Evertson then at the University of Texas, and David Berliner and William Tikunoff at the Far West Laboratory) were using methodology similar to ours to conduct research which ultimately led to important information and concepts relevant to teaching practice and research. Although this book focuses on our studies, it is important to recognize that a number of individuals (influenced in part by the earlier reviews of Dunkin and Biddle, 1974; Rosenshine and Furst, 1973; and by social conditions, including the interest of the National Institute of Education) were beginning to argue that little could be learned about teaching unless researchers were careful to determine that teachers behaved differently or had different effects on students prior to the collection of extensive data at great expense.

SAMPLE SELECTION

Over 100 third- and fourth-grade teachers were initially studied. The data unit for this study consisted of individual students' total mathematics scores on the Iowa Test of Basic Skills. The test scores from the fall of the third grade were used as prescores for the third grade and tests given in the fall of the fourth grade were used as postscores. Similarly, the tests administered in the fall of the fourth grade were used as prescores for the fourth grade and postscores were obtained by fall testing in the fifth grade. These data were compiled for fall testing in 1972, 1973, and 1974. A residual gain score was computed for each student on each subtest by using the student's total mathematics prescore subtest as a predictor variable. Residual gain scores were computed within years and within grade levels (third and fourth). Data for teachers were then compiled by computing a mean residual gain score (from the scores for all students in a particular teacher's class) for year one and year two.

The initial sample included almost all third- and fourth-grade teachers in the school district. The testing procedure employed in the district required early fall testing. Unfortunately, the school district tested at every other grade level so that it was impossible to construct residual scores from extant data without additional testing, which we undertook.

Fall testing offers advantages as well as disadvantages. It can be argued that fall testing may improve the objectivity of teacher-administered standardized achievement tests in some cases because teachers are testing previous instruction, not their own. However, procedurally this arrangement constitutes a major data preparation problem. As we mentioned above, to compute a third-grade teacher's residual mean score it was necessary to find an individual student's achievement score in that teacher's room in the fall, and also to find that student's score the following fall (that student, of course, may be in one of four or five fourth-grade rooms).

To further complicate matters, teachers may share students (with another teacher) for instruction in math or reading. Hence, even though a student's name appears on one teacher's Iowa achievement roster the student may or may not be instructed by that teacher during language arts or mathematics. Approximately one-half of the teachers in this sample shared students for instructional purposes (mathematics 45 percent; reading, 55 percent). Sharing in some cases involved the transfer of only two or three students; however, in some classes over a third of the students transferred. We provide this caveat (school records may not accurately associate teachers and students) so that others who contemplate research in this area will know that an adjustment of school records is a necessary part of such research.

Although we had anticipated that most of our data would be collected during mathematics, we hoped that it might be possible to identify many teachers who were relatively high in mathematics and reading across consecutive years. However, the general instability in the sample made it impossible to use multiple selection criteria/procedures across subject matter in any robust way. Even within the relatively stable teaching context of our sample, teachers' median year-to-year correlation across residual gain scores on all Iowa subtests was only .20. Teachers' average effectiveness (the mean score of their students on a particular subtest) clearly fluctuated from year to year.

We had also considered using students' reports of teacher rapport to identify teachers who had obtained high levels of student satisfaction over consecutive years. However, using the Rabinowitz and Rosenbaum Halo Scale (Rabinowitz and Rosenbaum, 1958), we found that the year-to-year stability of student affective responses (same teachers, different groups of students) was only .22. Several factors could explain the lack of stability of the achievement and affective assessments (e.g., homogeneous student population; the affective means were all high, and thus variance between teachers was quite restricted). Readers who are interested in a thorough discussion of these data can consult Good and Grouws (1975a, 1975b).

The question of *why* teachers are unstable from year to year is an interesting one. Perhaps subtle differences between student populations of classrooms (e.g., although the mean achievement level and the range of student ability appear similar across years, the presence of one or two students may make teaching one class more difficult than another) or events in teachers' personal lives (they have more time to plan for teaching in some years) account for fluctuation in teacher effects on student achievement. However, since the purpose of the present study was to study the behavior of teachers who had positive and stable effects on students' achievement, we did not address the issue of instability nor were we able to use multiple criteria to identify our teacher sample. Ultimately, we chose to study intensively only teachers who were stable and relatively high or low in their effects on students' average performance on the total mathematics score of the Iowa Test of Basic Skills.

We identified nine fourth-grade teachers who were relatively effective and stable on total math residual scores across two consecutive years (that is, they were in the top third of the sample across two years), and nine fourth-grade teachers who were relatively ineffective and stable for two consecutive years. During the year of observation, the high- and low-effective teachers maintained their relative patterns of achievement. Hence, these teachers were stable across three years.

PROCEDURES

Although our primary interest was in observing the 18 target teachers, observation data were collected in 41 classrooms to protect the identity of the relatively effective[1] and ineffective teachers. Approximately equal numbers of observations of mathematics lessons were made in all classrooms (6 to 7). Data were collected by two trained observers (both certified teachers) who worked full time and lived in the target city. Prior to the collection of observational data, the observers participated in a three-and-one-half–week training program that involved the coding of written transcripts, videotapes, and ongoing classrooms. Training was terminated when the two coders exhibited over 80 percent agreement on all coding categories. Each coder visited all 41 teachers and made about one-half of the observations obtained in each classroom. Furthermore, coders made all observations without knowledge of the teachers' levels of effectiveness.

During October, November, and early December, observational (process) data were collected in the rooms of the 41 participating teachers during mathematics instruction. These data were subsequently compared with classroom mean residual total mathematics scores on the Iowa Test of Basic Skills obtained during October and April of the same school year.

Four basic types of information were collected in the study. First, measures were taken to describe how mathematics instructional time was utilized. Besides serving descriptive purposes, these categories were designed to facilitate the testing of several hypotheses and resulted from previous experimental research on the utilization of time during mathematics instruction (e.g., the ratio of time spent in development vs. practice). The second set of codings consisted of low-inference descriptions of teacher-student interaction patterns. These data were collected with the Brophy-Good Dyadic System (1970). The third set of data was high-inference variables (describing general teacher behavior and managerial style) drawn from the work of Emmer (1973) and Kounin (1970). The fourth type of data coded were checklists that were used to describe materials and homework assignments (for more details see Appendix A and Good and Grouws, 1975a).

In presenting the results, positive findings will be emphasized and teaching behaviors will be interpreted as *clusters of behaviors*. This position is assumed for two reasons: (1) in order to communicate the results succinctly; and (2) because we do not believe that individual teaching acts constitute an appropriate way to conceptualize instructional behavior. Rather, any particular teaching act is meaningful only in relationship to other instructional behaviors that occur in the classroom.

---

1. As noted in Chapter 1, we defined *effectiveness* as teachers' ability to produce results on a standardized achievement test. We utilize the term *relatively effective* because different criteria may yield a different sample of teachers.

RESULTS

Process data presented in this chapter were analyzed with a one-way analysis of variance model[2] to see if there were behavioral differences between the nine teachers with the highest residual gains and the nine teachers who had the lowest scores.

Before discussing behavioral findings, it is useful to note one context finding concerning classroom organization. Teachers who had extreme effects (both positive and negative) used a large-group organizational structure. The 18 target teachers occasionally had two or three students assigned to individual work but they basically taught math to the class as a whole. Hence, organizational format did not predict achievement gain; the quality of instruction was the key difference between teachers who were obtaining relatively poor and good results.

Several of the teachers (but not the nine highs and nine lows) included in the observational study taught mathematics to two or three groups of students. Teachers who taught mathematics via group instruction generally were in the middle of the distribution in terms of average residual gain scores on total mathematics; teachers who taught the whole class consistently appeared at the top or bottom of the distribution. These data suggest that teaching the class as a large group is not categorically either a poor or good strategy. If a teacher possesses certain capabilities, it may be an excellent strategy; if not, large-group instruction may not work well. Subsequently, we have replicated this pattern of results in another large school district. A few teachers who got very high or very low scores used individual or small-group methods, but most such teachers generally were in the middle of the distribution.

Nor did more and less effective teachers differ significantly in the total amount of time they allocated for instruction. Rather, we found that how teachers used time (quality of teaching) was the important issue. What follows, therefore, is our characterization of the major ways in which relatively effective and ineffective teachers differed in their classroom behavior during mathematics instruction. To reiterate, data will be presented selectively here and interpreted in a general way.

The ability to make clear presentations seems to be one of the necessary skills for effective whole-class instruction. Our data indicated that high teachers regularly exceeded lows in clarity scores; that is, they generally introduced and explained material more clearly than lows. In whole-class settings high teachers asked more product questions (questions that demand a single correct answer) and appeared to "keep the ball rolling." However, when students did experience difficulty, highs were more likely

2. Data were also analyzed with other samples (all teachers; the top and bottom three) and with other statistical procedures (correlation techniques). The interested reader can consult Good and Grouws (1975a) for more technical information.

than lows to give process feedback (a response that not only provides a correct answer, but also suggests how that answer could be derived). In contrast, lows asked more process questions (a question that demands integration of facts, explanation) and gave less process feedback.

It seems that high teachers did not always focus upon process information; instead, they gave process feedback only when student responses indicated some error (i.e., highs appeared to move class instruction in a diagnostic cycle). Teachers who obtained better student performance were more willing to deal with student failure and provide related corrective feedback than teachers who obtained lower levels of student achievement.

Although not directly reflected in coded behavioral measures, classroom observers in the project reported that more effective teachers first taught the concept under consideration and used product questions to be sure that students had understood the concept. Then, if students demonstrated basic understanding of what had been presented, teachers were more likely to question students about the process and raise more general issues associated .with explanation and understanding. If students' responses to product questions did not reflect comprehension, effective teachers retaught a concept or idea. Hence, it appeared that effective teachers more actively monitored instruction and were willing to adjust their behavior as necessary to student performance.

Effective teachers also taught their classes as a whole a larger percent of the time,[3] taught more students, and spent somewhat more time on development (explaining the meaning of concepts) than did less effective teachers. In general, high-achievement teachers presented students with a clear focus of what was to be learned, provided developmental (process) feedback when it was needed, structured a common seatwork assignment for their classes (when students' performance suggested that they were ready for such work), and responded to individual students' needs for help.

Two other patterns differentiated the behavior of high and low teachers. Highs demanded more work and achievement from students (communicated higher performance expectations). For example, high teachers assigned homework more frequently than lows and highs also moved through the curriculum considerably faster than low teachers. Although less effective teachers spent more class time discussing homework, observers' logs indicated that effective teachers assigned homework more frequently than less effective teachers (48 percent vs. 38 percent), and more effective teachers were found to move through the curriculum more quickly (1.13 pages per day vs. .71 pages per day) than less effective teachers.

---

3. Although high and low teachers both used large-group instruction as a basic format for instruction, high teachers spent more time teaching the class as an intact group.

Effective teachers appeared to provide more immediate, nonevaluative, and task-relevant feedback than did less effective teachers. Furthermore, several behavioral measures consistently demonstrated that high teachers were approached by students more than teachers in low classrooms. Presumably, when students in high classrooms wanted information or evaluative input, they felt free to approach their teachers. Even when a teacher dealt with the entire class in a public format, students in high rooms participated through their own initiative. Students in these rooms asked teachers more questions, called out more answers, and proportionately were asked more open questions (questions which students indicated they wanted to answer; they raised their hands, etc.).

Although we do not have data to support this point directly, we suspect that students in effective teachers' classrooms were more willing to seek academic information from teachers because of the nonevaluative climate in these classrooms (the teachers expected good performance but did not overreact to success or failure), and because students knew that information presented by their teacher would be used that day in their seatwork and homework assignments. If they understood the teachers' presentation, then students would be able to complete subsequent work successfully. In less effective teachers' classrooms, the relationship between understanding material presented by the teacher and success on subsequent activities was less clear to students (in part because less effective teachers' presentations were relatively brief and vague).

In the classroom context in which we observed, encouraging frequent student-initiated behavior was a practical instructional technique. That is, middle-class students who possess basic skills and have been given a clear explanation of the learning task should be allowed to approach the teacher as they need help. In a different context (low-aptitude students, teacher does not provide a clear structure for work), frequent student-initiated behavior may indicate confusion or managerial problems, and be associated with lower student achievement.

The data demonstrate that high praise rates do not necessarily enhance learning. Indeed, in this study praise was negatively associated with both achievement and climate.[4] High teachers consistently praised less than low teachers. However, despite their generally high praise rates, low teachers were much less likely than highs to praise individual students who approached them about academic work. Our data indicated that low teachers were more likely to go to students (rather than being approached

4. One interesting, and to some extent contradictory, finding is that during the initial study of over 100 teachers there was no correlation between teachers' mean climate scores (affect) and students' residual mean performance in mathematics. During the observation stage the correlation between these two product measures was .50. Among the explanations we have considered for this anomaly is that in the initial sample students were reacting to teachers and schools generally, but during the observational study students were responding to the more specific context of mathematics.

by them), a strategy that proved to be ineffective in this study. This finding also has to be interpreted in the context of the coding definition we used for praise. Had we used more stringent criteria (e.g., coding only appropriate or contingent praise) different conclusions might have been reached about the relationship between praise and achievement (see Brophy, 1982).

Still, it was clear in this study that high teachers were basically non-evaluative. They did not criticize or praise academic work as frequently as low teachers. The evaluative stance of lows, along with their high rates of approaching students, may have interfered with learning progress and may also have created a more negative climate with more tension. High classrooms were regularly described more favorably by students, despite the fact that high teachers did not praise much. Although observers rated the effective teachers as more task oriented, they also observed these classrooms as more pleasant than less effective teachers' classrooms. Hence, more effective teachers' classes were viewed as relaxed, but task oriented.

Low teachers seemed to have more frequent managerial problems than highs. The argument that less effective teachers had more discipline problems is based upon the finding that these teachers issued many more behavioral warnings, criticized students more often, and provided more alerting and accountability messages to students than did the more effective teachers.

The process through which teachers achieve well-managed classrooms is less clear. By inference, it appears that teachers who often use alerting or accountability behavior are ineffective (possibly because they are only responding to the misbehavior). Similarly, high praise and high criticism rates probably are associated with managerial problems because they are used after misbehavior occurs. It may be that by communicating high performance expectations, structuring clear learning goals, and providing students with immediate, helpful feedback, teachers prevent most management problems.

We have discussed some of the behavioral findings that tended to describe teachers who were obtaining relatively high and low levels of student achievement. It is also of interest to describe coders' general impressions of the teachers. Regrettably, teachers are often evaluated on the basis of one or two general observational visits by a supervisor. We wondered to what extend coders' impressions of teachers' abilities were correlated with student achievement.

After the project was completed, the two coders were asked to rank the 41 teachers in terms of perceived effectiveness in producing student learning gains. At this point the coders did not have access to the systematic records they had collected nor any statistical analysis or summary of these data. They were asked to give their general impressions of teachers' effectiveness. The coders were generally accurate in identifying the relatively ineffective teachers; however, many of the high teachers were described as performing at a medium level of effectiveness and some teachers who appeared to have average effects on student achievement were described as high. Furthermore, the two coders misclassified different teachers. Such findings suggest that "ineffective" teaching is easier to distinguish from average teaching than is "effective" training.

We infer from the coders' inability to accurately identify teachers who were getting the best gains that effective teaching is not necessarily one or two skills or behaviors which may be observed in infrequent visits to a classroom. Rather, such teaching is probably the accumulation of experiences and information which teachers provide for their classes that over time create meaning and understanding for students. In describing and evaluating teachers, some supervisors/coders may be impressed by instructional techniques or classroom demonstrations that interest them (but perhaps are not relevant to students' needs or perceptions) and/or neglect subtle but important teacher behaviors (e.g., assessing student comprehension of knowledge).

The data, discussion with coders, and our reflections upon both types of information led us to conclude that effective teaching (within the context in which we studied it) was best conceptualized as a system of instruction: (1) instructional activity is initiated and reviewed in the context of meaning; (2) students are prepared for each lesson stage to enhance involvement and to minimize errors; (3) the principles of distributed and successful practice are built into the program; (4) active teaching is demanded, especially in the developmental portion of the lesson (when the teacher explains the concept that is being studied, its importance, and so on).

DIFFERENTIAL ACHIEVEMENT GAINS

Why were some teachers in this study relatively more effective in helping students to master mathematics skills (as measured by the mean residual score on the Iowa Test of Basic Skills)? Our data address an important related issue: do effective teachers appear to systematically get their achievement gains from students at a particular achievement level? These data were drawn from the data base described earlier in this chapter (i.e., 103 third- and fourth-grade teachers from a large, middle-class school district. The descriptions that follow are brief summaries and the reader who wishes more technical detail is referred to Good and Beckerman, 1978; Good and Grouws, 1979).

For the comparison, 103 third- and fourth-grade teachers, within grade level, were initially assigned to high, middle, and low groups on the basis of their residual gain scores. Students were assigned to high-, middle-, and low-aptitude groups according to their scores on the Cognitive Abilities Test. If teachers did not have four students in each cell they were dropped from the analysis. Applying this criterion reduced the number of teachers from 103 to 81 (40 third-grade teachers and 41 fourth-grade teachers). Teachers who were excluded from the analysis came from all three teaching levels.

Two major analyses were then performed. First, all third- and fourth-grade students were divided into three equal groups according to aptitude scores. Students in the top third were assigned to the high group and so forth. This analysis was called the *absolute analysis*. The second analysis involved the assignment of students into thirds on the basis of their aptitude ranks within their own mathematics classes. This analysis was termed the *relative analysis*. A 3 (teacher competence) × 3 (student aptitude or achievement level) analysis of variance with repeated measures was computed, with student residualized gain scores (on the Iowa Test of Basic Skills) as the dependent measures for both relative and absolute analyses.

Effective teachers, as a group, did not appear to achieve their results because of the performance of one student-aptitude group. Similarly, ineffective teachers, as a group, are not especially ineffective with one

TABLE 2.1   Anova: Differences in Residual Achievement for Three Levels of Teacher Competence and Three Levels of Student Aptitude—Grade 3 (Relative Analysis)

Source of Variation	SS	F Value	df	Significance
Between 3 levels of teacher competence	272	45.80	2	p<.001
Error	112		38	
Between 3 levels of student aptitude	169	18.10	2	p<.001
Between teacher competence (teacher nested) by student aptitude	28	1.48	4	.21 N.S.
Error	356		76	

Anova: Differences in Residual Achievement for Three Levels of Teacher Competence and Three Levels of Student Aptitude—Grade 4 (Relative Analysis)

Source of Variation	SS	F Value	df	Significance
Between 3 levels of teacher competence	383	42.16	2	p<.001
Error	199		37	
Between 3 levels of student aptitude	197	21.71	2	p<.001
Between teacher competence (teacher nested) by student aptitude	7.46	.41	4	.80 N.S.
Error	336		74	

aptitude group. Highly effective teachers as a group do a better job with students at each level; less effective teachers as a group do a relatively poorer job with all students. Although it was possible to identify a few teachers within each effectiveness group who appeared to do an especially good or poor job with a particular student group, there were no consistent patterns. Summaries of the variance tests are presented in Tables 2.1 and 2.2 and the means associated with these tests appear in Table 2.3.

Within the context of this study (e.g., a middle-class school district) and with particular student achievement data (means and variances), teacher effectiveness and student aptitude were not found to interact in any systematic way. Unfortunately, the pattern of these results did not provide clues about how to design the observational procedures for the experimental study nor did it offer useful information about how the treatment program should be designed for particular students. For example, had we found that effective teachers achieved their status because of the performance of certain types of students, it would have been important to determine how teachers interacted with these students. Considering the general nature of our definition of teacher effectiveness in this study, there were no compelling reasons to study individual students during the initial field experiment or to build special features for particular types of students.

TABLE 2.2   Anova: Differences in Residual Achievement for Three Levels of Teacher Competence and Three Levels of Student Aptitude—Grade 3 (Absolute Analysis)

Source of Variation	*SS*	*F Value*	*df*	*Significance*
Between 3 levels of teacher competence	291	39.02	2	p<.001
Error	142		38	
Between 3 levels of student aptitude	135	14.05	2	p<.001
Between teacher competence (teacher nested) by student aptitude	36	1.88	4	.12 N.S.
Error	365		76	

Anova: Differences in Residual Achievement for Three Levels of Teacher Competence and Three Levels of Student Aptitude—Grade 4 (Absolute Analysis)

Source of Variation	*SS*	*F Value*	*df*	*Significance*
Between 3 levels of teacher competence	373	41.74	2	p<.001
Error	165		37	
Between 3 levels of student aptitude	196	15.01	2	p<.001
Between teacher competence (teacher nested) by student aptitude	56.68	2.24	4	.07 N.S.
Error	483		74	

TABLE 2.3 Main Effect Means from the Relative and Absolute Analyses for Grade 3 and Grade 4

	Grade 3 Relative Analysis		Grade 4 Relative Analysis		Grade 3 Absolute Analysis		Grade 4 Absolute Analysis	
	Teacher Competence Level	Residual Mean	Teacher Competence Level	Residual Mean	Teacher Competence Level	Residual Mean	Teacher Competence Level	Residual Mean
	Low	-1.96	Low	-2.35	Low	-1.97	Low	-2.23
	Middle	0.01	Middle	-0.01	Middle	0.05	Middle	-0.38
	High	1.69	High	2.08	High	1.81	High	2.13
	Student Aptitude Level	Residual Mean	Student Aptitude Level	Residual Mean	Student Aptitude Level	Residual Mean	Student Aptitude Level	Residual Mean
	Low	-1.59	Low	-1.77	Low	-1.47	Low	-1.79
	Middle	-0.05	Middle	0.12	Middle	0.17	Middle	-0.07
	High	1.25	High	1.37	High	1.04	High	1.37

CONCLUSION

We were able to identify teachers who consistently obtained more mean classroom achievement than did other teachers who were teaching similar students under like circumstances. Subsequent classroom observation demonstrated that sets of classroom behaviors differentiated relatively effective from relatively ineffective teachers. In general, effective fourth-grade mathematics teachers, in contrast to less effective teachers: presented information more actively and clearly; were task focused—spent most of the instructional period on mathematics, not socialization; were basically nonevaluative and created a relatively relaxed and pleasant learning environment—offered comparatively little praise or criticism; expressed higher achievement expectations such as more homework, a faster pace, a more alert environment; and had to issue fewer behavior statements to students. We were pleased that it was possible to isolate patterns of instructional behavior which were associated with student achievement. However, we were well aware of the possibility that many factors other than the behaviors we had observed in high-achievement classrooms might be responsible for the higher achievement of students. For example, perhaps more effective teachers plan more thoroughly and because of this they are more task focused, assign more homework, etc.

Correlational findings do not lead to direct statements about behaviors teachers should utilize in classrooms. That relatively ineffective teachers were rated by observers as being unclear may be partially related to the fact that they did not spend enough time in the developmental stage of the lesson and/or because they did not listen and respond appropriately to students' responses early in the lesson, not just because of teachers' weak skills in presenting material.

Despite these and many other problems associated with translating research findings into practice,[5] our initial findings were sufficiently interesting to us that we decided to design a teaching program in order to test the ideas in a field experiment. In designing our treatment program, we were especially aware of our naturalistic findings; however, as Chapter 3 will show, we also based our program on the work of other process-product researchers, extant experimental work in mathematics education, our own hunches (as well as those of project staff members), and insights stimulated by classroom observation.

5. In this study we interpreted the findings only as the teacher reacted to the class, using only linear models, and with no direct measurement of teachers' and students' perceptions of classroom processes. These and other issues associated with teacher effectiveness research will be discussed in Chapter 10.

3

Building a Treatment Program

We were pleased with the fact that some consistent differences could be found between relatively effective and ineffective mathematics teachers. However, at that point we only had a description of how more and less effective teachers (in our sample) behaved differently. We wanted to integrate our findings with those of others and to determine whether teachers could be taught these effective teaching behaviors and to see if such instruction and training could be used to improve the mathematics performance of students.

We were aware that our initial results were only correlational data and that they did not necessarily imply that differences between high- and low-achieving teachers caused student achievement. It could very well be that behaviors not studied in our observational research are more directly related to achievement (for example, more effective teachers plan more thoroughly and because of this, they are more task focused, assign more homework), or that these teachers taught more actively because they had more energy or because of other personality characteristics. We felt that it was important to determine whether a more direct association could be established between the behaviors which were identified in our observational, naturalistic study and student achievement.

In writing the teacher training materials, our earlier naturalistic findings were integrated with the recent naturalistic research of others and with existing experimental research in mathematics education, and translated into an instructional program. Some of the variables we tested in our experimental program did not come directly from teaching behaviors measured in our earlier studies, but were instead based upon what observers had seen in classrooms or in research conducted elsewhere. Although the treatment program was consistent with our earlier naturalistic findings, it was broader than the findings per se and the findings had to be translated. Emphasis is placed upon the term *translation* because correlational findings (or other forms of research findings) do not provide direct implications for action.

For example, we found in our naturalistic study that the correlation between development and achievement was negative. However, in the case study of only high and low effective teachers, we found that highs spent more time on development than did lows (although relatively little time was spent on development by both groups of teachers). In this case because of observers' reports, impressive experimental data collected by others (see Chapter 9) and because of our own professional judgments we made development a major treatment variable. In general, in building the treatment program most reliance was placed on the data differentiating high and low teachers and many of the decisions about treatment variables were based upon these data directly. Still, we want to emphasize that translation of findings into teaching recommendations involves considerable judgment and interpretation.

Because of our interest in studying the program and in determining whether these strategies would have impact upon students' learning, we encouraged teachers to follow the recommended time distributions closely. We knew that the amount of time one should spend on development varies with the type of material being presented and with the purpose of the lesson (e.g., an introductory or a review lesson). Several qualifications will be presented later in the book but in this chapter we want to present the program in the same way it was presented to the treatment teachers. To reiterate, for research purposes we urged teachers to follow the program as closely as possible. The program we asked teachers to use is presented below.

MISSOURI MATHEMATICS GENERAL PROGRAM MANUAL

We believe it is possible to improve student performance in mathematics in important ways. We look forward to your help and cooperation in implementing the program that we have discussed at the workshop and which is outlined again in the material that follows. Through your efforts we believe a significant difference in student performance will be made.

We do not believe that any single teacher behavior will make a critical difference in student learning, but we do feel that several behaviors in combination can make a major impact. In the material that follows, we present a system of instruction that, if followed daily, will enhance student learning.

In general, we feel that the plan should be followed each day. However, we also realize that special circumstances will force you to modify the plan on occasion. Still, it is important that you follow the daily plan as frequently as you can.

TABLE 3.1   Summary of Key Instructional Behaviors*

Daily Review (first 8 minutes except Mondays)
1. Review the concepts and skills associated with the homework
2. Collect and deal with homework assignments
3. Ask several mental computation exercises

Development (about 20 minutes)
1. Briefly focus on prerequisite skills and concepts
2. Focus on meaning and promoting student understanding by using lively explanations, demonstrations, process explanations, illustrations, etc.
3. Assess student comprehension
   a. Using process/product questions (active interaction)
   b. Using controlled practice
4. Repeat and elaborate on the meaning portion as necessary

Seatwork (about 15 minutes)
1. Provide uninterrupted successful practice
2. Momentum—keep the ball rolling—get everyone involved, then sustain involvement
3. Alerting—let students know their work will be checked at end of period
4. Accountability—check the students' work

Homework Assignment
1. Assigning on a regular basis at the end of each math class except Fridays
2. Should involve about 15 minutes of work to be done at home
3. Should include one or two review problems

Special Reviews
1. Weekly review/maintenance
   a. Conduct during the first 20 minutes each Monday
   b. Focus on skills and concepts covered during the previous week
2. Monthly review/maintenance
   a. Conduct every fourth Monday
   b. Focus on skills and concepts covered since the last monthly review

---

* Definitions of all terms and detailed descriptions of teaching requests will follow.

For purposes of clarity, we will discuss each part of the teaching program separately. However, once again we want to emphasize that the program works when all parts are present. To maximize your opportunity for obtaining a clear picture of the instructional program, the program is summarized in Table 3.1. The rationale for each part and how the pieces fit together will be discussed at a later point in the handbook.

DEVELOPMENT

Variable Description

The developmental portion of the mathematics period is that part of the lesson devoted to establishing comprehension of skills and concepts. It should be viewed as a continuum which runs from developing understanding to allowing for meaningful practice in a controlled setting. During all stages of the developmental portion of the lesson, both ends of the continuum may be present to some degree. However, in general, the comprehension emphasis with very little practice will come at the initial part of the lesson, then toward the middle of the lesson, practice with process feedback from the teacher will become quite prominent, and finally in the latter portion of the lesson there will be controlled practice with meaningful explanations given as necessary.

The role of the teacher in the first part of the lesson, the comprehension phase, is to use instructional strategies that help students understand clearly the material being studied that day. In this portion of the lesson emphasis is placed upon comprehension rather than rote memorization. Activities which often focus on comprehension include teacher explanations and demonstrations, and they may include use of manipulative materials to demonstrate processes and ideas, use of concrete examples in order to abstract common features, making comparisons and searching for patterns, and class discussions.

During the middle portion of the lesson the number of questions posed to students may increase as the teacher begins to assess comprehension and provides them an opportunity to model processes already demonstrated, and to verbalize the understanding they have developed. During this phase of the lesson the teacher may decide that further explanations and demonstrations are necessary or decide that controlled practice is appropriate since students seem to understand what they are doing.

In the controlled practice phase of the lesson the emphasis is on increasing proficiency; that is, increasing speed and accuracy. However, meaningful feedback is still given as necessary or requested.

Problem

Many problems arise in math classes in which teachers give too little attention to development. Students exposed to such teaching frequently attempt to memorize rules for doing things and concentrate on mechanical skills. These rules have no meaning for the student (because developmental work was not done) and, thus, they are easily forgotten, especially when new sets of rules are "learned." When students do not understand what they are doing, each new problem causes them great difficulty. Often the comment, "We haven't done any of these before," is heard. When students learn without understanding, the ability to transfer skills to new situations is greatly reduced. Other negative results such as the inability to detect absurd answers and loss of self-confidence also occur. Thus, there are many compelling reasons to include a large measure of developmental work in mathematics lessons.

Teaching Practice

Initially, the teacher should focus briefly upon prerequisite skills that students may need for the lesson. Then the major aspect of the meaning portion of the development lesson occurs: active demonstration of the concept, idea, and/or skill that is being focused upon in the lesson, etc. Teachers need to demonstrate the process actively, so that students can comprehend the learning goal. You need to be cautious about moving too quickly into the assignment of problems and practice without providing students with an adequate conceptual orientation.

After the active demonstration and explanation by the teacher (and we recommend that ten minutes minimum be spent on this), the teacher should begin to *assess student comprehension*. There are two primary ways to do this. First, teachers may ask oral questions. In general, we recommend that teachers generally ask brief product-oriented questions. Product questions are questions that assess whether or not the student can produce the correct answer (see Appendix 3.A for a complete description). Teachers can maintain an emphasis upon meaning by frequently providing process explanations themselves after students respond ("Yes, Tina, that's right because . . .").

The second way that teachers can assess student comprehension is by having students work practice problems. However, it is important to recognize that the role of a practice problem in this stage of the lesson is not to build up student speed and accuracy per se, but rather, the goal is to allow teachers to assess student comprehension. Hence, the assignment of problems in this stage should be limited to a single, brief problem followed by teacher assessment and explanation and then the provision of another brief problem assignment. In general, this stage of the lesson can be completed in three to five minutes.

If your questions or assigned problems reflect a moderate degree of student difficulty, then you should repeat the meaning portion of the lesson. If possible, use different examples; however, if this is not possible, verbatim repetition of the initial portion of the lesson is better than to proceed to controlled practice and seatwork when students are confused. Such a situation guarantees that students will practice errors.

If assessment of student comprehension is largely satisfactory, then teachers should proceed to the controlled practice portion of the development lesson. Now, the teacher provides opportunity for students to develop increased speed, accuracy, and proficiency in completing problems of a specific type. However, the practice is still heavily controlled (unlike seatwork practice, which will be discussed in the following section).

During controlled practice, teachers should assign only one or two problems at a time. *Students should not be asked to work longer than a minute without feedback about the correctness of their responses.* The reason for this is that during the controlled part of the lesson the teacher is still trying to identify and correct any student misunderstanding. Too often many students are left to watch while a few students demonstrate a problem on the board. A great deal of practice time is lost this way and often the involvement of some students in the lesson (momentum) is lost as they become engaged in side conversations and distractions.

During controlled practice exercises, teacher accountability and alerting should be immediate and continuous. By alerting, we mean teacher behaviors that remind students that they should be doing work and that it will be checked. For example, if the teacher sends three or four students to the board to demonstrate the problem that students have just worked at their desks, the teacher might say, "Now the rest of you do these two new problems at your desk and I'll check them in a minute." Such teacher behavior maintains student momentum. Instead of watching classmates write on the board, they have their own work to do and they are alerted to the fact that they will be held responsible for the work.

By *accountability* (more on this when we describe the seatwork portion of the lesson) we mean the actual checking of student responses. For example, while students put their work on the board, the teacher could look at the work of students who remain at their desks and check the problems that they were to have completed. Furthermore, the teacher can call on students to provide answers to practice problems, etc. Through such procedures, the teacher is able to assess when students are prepared to move to the seatwork portion of the lesson, where they have a longer block of time for uninterrupted practice. A final important characteristic of the controlled seatwork portion of the lesson is that the practice is done in the context of meaning (e.g., the teacher is frequently providing process explanations: "Yes, that's right because. . . ."). Although the teacher is beginning to work for speed and accuracy, some attention is still being paid to students' understanding of the concepts, ideas, and skills that are being developed.

In summary, the development part of the lesson calls for the following teacher behavior:

1. Review briefly and/or identify prerequisite skills.
2. Focus upon the development of meaning and comprehension using active demonstration and teacher explanation.
3. Assess student comprehension (ask questions/work on supervised practice).
4. Repeat meaning portion of the lesson as necessary (using different examples and explanations if possible).
5. Provide practice opportunities for students.
   a. Practice should be short (one or two problems at a time).
   b. Students should be held responsible for assigned practice problems.

    c. Practice should be performed in a meaningful context (teacher provides frequent process explanations).

    d. When success rate is high, move students into seatwork portion of the lesson, where students have an opportunity for uninterrupted practice.

SEATWORK

Variable Description

Seatwork refers to practice work that students complete individually at their desks. Since seatwork practice follows the controlled practice part of the development lesson, students should know the purpose of assigned problems and how to do them when they begin to work. The role of seatwork practice is both important and easy to describe. Seatwork assignments allow students to practice, on their own, problems and principles that you have just actively taught. Seatwork provides students with an opportunity for immediate and successful practice. This practice experience allows students to achieve increased proficiency and to consolidate learning. New material or review work should not be assigned during the seatwork portion of the lesson.

Problem

Often a great deal of time is wasted when students work on problems individually. Indeed, research has consistently shown that students show less involvement (amount of time that students actually spend working on problems) during the seatwork portion of the lesson than during the active teaching portion of the lesson. Too often teachers stop active supervision after they make the seatwork assignment. Two of the more common ways that teachers stop supervision are by doing deskwork, grading, or by providing extended feedback to a single student (before all students are working on the task). Such behavior virtually guarantees that teachers cannot provide the type of supervision that students need if they are to begin to work productively. The first teaching task is to get students started on the seatwork. Often students do not use seatwork time productively simply because the teacher does not obtain their attention initially.

    In addition to the problem of not "demanding" students to start work, some teachers create a problem by moving from the development portion of the lesson to seatwork with such abruptness that it is not surprising that students do not begin to work immediately (e.g., four students spring to the pencil sharpener, two students search for materials, and three students begin a private conversation). Momentum (student attention and involvement) needs to be maintained throughout all stages of the lesson. When momentum is lost, students are apt to take a psychological break and once it is lost, it is difficult to "recapture." Teachers who end the development portion of the lesson with a controlled practice segment have done much to ease the transition from the group lesson to individual seatwork.

Teaching Request

Given that the role of seatwork is to provide opportunity for successful practice, we recommend that about 10 to 15 minutes each day be allotted for seatwork. Ten to 15 minutes allow sufficient time for students to work enough problems to achieve increased proficiency but not so long as to bring about boredom, lack of task involvement, and the behavioral problems that soon follow when students are bored or frustrated. Frustration should be minimal in seatwork activity because the problems students are asked to do are a direct extension of the development part of the lesson. If practice time does not exceed 15 minutes, few students are likely to be bored.

The number of problems assigned should take most students only 15 minutes to complete. Hence, approximately 75 percent of your students should be able to complete the work within the allotted 15 minutes. In making the seatwork assignment, emphasis should be placed upon accurately working as many problems as possible within the allotted time. Low achievers who remain on task and do accurate work have done well and should know that they have done well. That is, the criterion to communicate to students is to keep working and do as many problems accurately as they can.

To help optimize the effectiveness of seatwork, three general principles should be observed. The first principle, momentum, has already been discussed indirectly. By momentum we mean keeping the ball rolling without any sharp break in teaching activity and in student involvement. Teachers can achieve momentum by ending the development portion of the lesson by working problems similar to the ones that students are asked to work individually and by starting students on individual work with a simple and direct statement. "We've worked problems 1 and 3. Now, individually, at your desks do problems 5 through 15. Work as many problems as you can, and we'll check our work in 15 minutes. Remember, doing the problem correctly is more important than speed." Following such a statement, you should actively monitor all students. Before providing feedback to individual students, make sure all students are engaged in the seatwork.

If some students do not begin working immediately, walk to their desks and if your physical presence doesn't initiate student work as it usually will, then quietly say something like, "Frank, it's time to do the problems." After all students are working on the problems (the ball is rolling), you can then attend to the needs of individual students. In general, students should get immediate feedback and help when it is needed. Thus, it is usually reasonable to allow students to approach you when they have a question or problem. However, when presenting feedback to individual students, keep in mind the general principle of momentum. You have to provide feedback and conditions that allow most students to stay on task (keep working). Hence, it is not advisable to continue to provide lengthy feedback to an individual while several students are waiting for teacher feedback before they can continue to work.

Alerting is a second principle to observe during seatwork. Alerting behaviors tell students that they will be held accountable for their work. Often students engage in off-task behavior because they are not alerted to the fact that they will have their work checked at a specific point in time. If students are assigned seatwork that won't be checked until the following day (or when it is not checked at all), students are not likely to be highly engaged in seatwork. A statement like, "We'll check the work at the end of the period," alerts students to the fact that there is reason to engage in productive work immediately. A statement at the beginning of the seatwork is sufficient. Repeated statements are apt to interfere with students' work concentration. Public announcements should not occur during seatwork. Once you have students working it doesn't make sense to distract them.

Accountability is the third principle to observe during seatwork. Alerting, as we noted, is a signal to students that their work will be checked. Accountability is the actual checking of the work. It is important that your accountability efforts do not interrupt the seatwork behavior of students. During the controlled practice part of the lesson (see development section), accountability is immediate. However, during the seatwork portion of the lesson, students are to be working more independently and those students capable of doing the work need time for uninterrupted practice. Public accountability needs to be delayed until the end of the lesson. A teacher's public questions during this stage of the lesson are very disruptive. For example, when the teacher asks a public question (e.g., "How many of you have done the first four problems?" "What's the answer to the second problem?") all students stop work, and once momentum is lost, some students will take much time before resuming their work. Furthermore, questions like, "How many of you have finished the first four?" may make students anxious and distract them from task behavior if they have not worked the first four problems. Occasionally, you may need to stop seatwork practice to correct a common misunderstanding. In general, these errors should be corrected during the development (controlled practice) phase of the lesson. Public statements (except for necessary behavioral management) should be avoided. If most students are not ready for seatwork practice, then you should stay in the controlled practice part of the lesson. Such behavior will help students develop the following attitude toward seatwork: "I can do the problems and now it is time for me to apply myself."

Perhaps the most direct and easiest way to hold students publicly accountable without disrupting seatwork is to call on individual students at the end of the lesson. Checking students' work at the end of the period also provides the teacher with a chance to spot any systematic mistakes that students are making and to correct those misunderstandings. Hence, when your students are assigned their homework, conditions should be set so that the homework provides for additional and relatively successful practice.

Specifically, we are asking you to get student involvement immediately after making a seatwork assignment. Continue to monitor and supervise all students until they are engaged in assigned work (the first minute or two). Early in the seatwork period (the first three to five minutes), be available for students when they need feedback. Toward the end of the seatwork period, try to get to the desks of some low achievers to see if they are making any systematic errors and to provide feedback as necessary. At the very end of the seatwork period, hold students accountable for their work by asking individual students to give the answer to a few of the assigned problems. This checking of answers should be very rapid and you need only check three or four of the problems (check one or two problems at the first, in the middle, and at the end of the assigned work). If misunderstandings are corrected here, the homework should be a successful practice experience for most students.

When conducting the review of seatwork, it is generally advisable to call on low-achievement students to provide answers only to the first few problems assigned so as not to frustrate them for failure to complete all problems, but be sure to increase seatwork expectations for these students as the year progresses.

Finally, all seatwork should be collected. This helps encourage students to work productively because they know that they are held accountable for the work assigned during seatwork. Because of the way teachers have used seatwork in the past, many students have built up the expectation that seatwork is a time to relax and waste time. Taking up the seatwork will help students to adjust to the expectation that seatwork is a time to apply themselves and to see if they can do the type of problems which will be assigned as homework. Although there is no compelling reason to grade seatwork, it is important to examine the papers to see if students are using seatwork time appropriately. If a student's work is unduly incomplete, impossible to read, etc., it would be important to mention this to the student so that he or she knows that you care about seatwork performance.

After the seatwork is collected, the homework assignment is made. Delaying the assignment of homework helps to insure that students will do the work at a later point in time, hence, building distributed (repeated) practice into the mathematics programs. Research has consistently shown the superiority of distributed practice over mass practice in helping students to master and retain new concepts and skills.

HOMEWORK

Variable Description

Mathematics homework is written work done by students outside the mathematics class period. It is usually done at home; thus, it is distinctly different from seatwork, which is done within mathematics class time.

Problem

The emphasis on homework in schools over the years has varied considerably. Unfortunately, homework has been misused frequently. Sometimes the assignments were so long that students became bored and careless when working the assigned problems. No doubt some students' dislike for mathematics is in part associated with these lengthy assignments. The instructional value of long homework assignments is very questionable. If students make errors on the first few problems of the assignment, then by the end of the assignment they may have become more proficient in making those errors!

Other situations in which homework has not been used to its full potential are plentiful. In some schools homework is never given or so few problems are assigned that an excellent opportunity for distributed practice is wasted. Another undesirable situation occurs when homework is given primarily to please parents but without much attention to selecting problems and assignments that will foster progress toward important objectives. *But perhaps the most devastating misuse of homework is when children are assigned problems for which inadequate background has been developed in class.* While long assignments often lead to frustration, this latter situation always leads to frustration and negative attitudes toward the mathematics class.

Another situation which detracts from the value of homework assignments happens when the teacher fails to stress the importance and value of the problems assigned. This can be done directly by not commenting on the importance of assignments or indirectly by not scoring or collecting assignments.

In spite of these misuses of homework, homework can be an important part of mathematics learning if certain guidelines are followed. Research suggests that giving homework to students on a regular basis may increase achievement and improve attitudes toward mathematics. Short assignments have been found to be most effective and some variety in the type of homework is helpful. Also, if a teacher gives importance to the homework through oral comments and by scoring papers regularly, then students frequently respond by completing their assignments with greater care.

Teaching Request

Because of the important role that homework can play in improving student performance in mathematics, we would like to have you do the following during the study:

1. At the very end of the math class period on Monday through Thursday, give a homework assignment which is due at the beginning of the class period the following day.
2. Each assignment should require about 15 minutes of outside class time. Within this time frame, assignments will probably average about eight problems per day depending on the kinds of problems being assigned. A typical assignment is shown in Appendix 3.B.
3. The primary focus for an assignment should be on the major ideas discussed in class that day. Also each assignment given on Tuesday and Wednesday should include one or two review problems from the current week's work.
4. Each assignment given on Thursday should be primarily devoted to review problems from the current week's work. In order for sufficient practice to be given on the material discussed on Thursday, this assignment will be a bit longer than assignments for other days and will probably take about 20 minutes for most students to complete.
5. Typically, each assignment should be scored (number correct) by another student. Papers should then be returned to their owners for brief examination. Finally, papers should be passed forward so that the scores can be recorded in the grade book.
6. The assignments given should be recorded daily in the teacher's log.

The short homework assignments complement seatwork by distributing practice over time without putting undue time pressure on students. Short assignments help hold student interest; adding variety to assignments is also helpful. This can be done by embedding the problems to be worked in different formats such as games, puzzles, codes, and so on. Appendix 3.C illustrates this idea. Another component of variety might be to have students check their work. Multiplication problems can be checked by doing division, addition problems by doing subtraction, and so on. Variety can also be introduced by giving differentiated assignments. For example, some students could be given ten easy problems, while other students are given six problems of a more difficult nature.

The scoring and recording of grades on all homework assignments are designed to emphasize the importance of homework and to provide regular feedback to students and teachers regarding progress being made by each student. It is important to realize that there are a number of efficient ways to score homework other than the teacher's going through the papers individually. For instance, students can exchange papers or score their own papers. Either of these procedures is improved if students are expected to have their homework completed and ready to be scored at the very beginning of math time. Efficiency is also improved if answers are prepared in advance by the teacher in written form (transparency, blackboard) and then shown to the students. Otherwise, the teacher may need to orally repeat each answer a large number of times.

Explanations and reteaching the homework must be somewhat limited if adequate time for discussion and practice of new material is to be available. This should not cause too much difficulty because most student difficulties and errors should have been remediated prior to the seatwork of the previous day.

A good strategy may be to have children quickly exchange and score papers, then have children indicate by raising their hands how many missed problems #1, #2, and so on. Then you can rapidly work the one or two problems that caused students the most difficulty. Since there are usually only a small number of homework problems to be checked and discussed, this part of the lesson should be easily completed in two minutes. Finally, note that any reteaching that is not completed can be done during the weekly review that is discussed in the next section.

In the rare event that the checking of homework reveals numerous student errors, you should reteach the previous day's lesson beginning with development, then controlled practice, then seatwork, and finally a homework assignment on the same material. Under these circumstances you should not try to cover new material due to the very limited amount of time available to develop the new ideas.

You are requested to personally score the homework that is assigned on Thursdays. There are two reasons for this. First, the information gathered from this homework is to be used to structure the weekly review each Monday. Second, the focus of student scoring is of necessity on answers rather than kinds of errors being made. It is very important, however, that regular attention be given to the procedures and processes that students are using. This is especially true when they are making errors!

In connection with the scoring of Thursday's work, each student's paper should be analyzed for systematic error patterns. Systematic error patterns refer to incorrect procedures which are consistently used on a wide range of problems. In two-digit multiplication problems, for example, a student might consistently forget to "carry" the tens digit from the initial multiplication of the unit digits. According to recent research, such errors are much more common than was once realized and, thus, spending time examining homework with them in mind can be very helpful in remediating some students' difficulties with mathematics. Further examples of common computational error patterns can be found in Appendix 3.D. Since the particular errors you find probably will not be associated with groups of students, the remediation of such errors is usually best done on a one-to-one basis.

Homework is an important component of this program and since both students and teachers devote a considerable amount of time to it, it is recommended that homework count at least 25 percent of each student's math grade and that this information be communicated to them.

Parents are interested in and should be informed about what is happening in school. Therefore, it is recommended that an explanation of the homework policy to be followed during the study be sent home to parents. A letter which could be duplicated and used for this purpose can be found in Appendix 3.E.

Homework is explicitly related to each of the other components of the study in a number of ways. With an increase in development time, it provides an opportunity to supplement the practice part of the lesson. It is structured such that practice is distributed over time and students have an opportunity to correct difficulties encountered in seatwork. The homework provides important information for structuring the specific details to be covered in the review component. It is also related to the pacing variable in that it allows some necessary work to be done outside of the time regularly scheduled for math.

SPECIAL REVIEW/MAINTENANCE[1]

Variable Description

Children forget. It is imperative, therefore, that ideas be reviewed and skills maintained on a systematic basis in elementary school mathematics. Reviewing ideas may involve the teacher's stating and explaining properties, definitions, and generalizations and the students' recalling the appropriate term or name. These roles occasionally may be reversed

---

1. The review discussed here is in addition to the brief (one to four minute) daily review that we will discuss later in the handbook.

(where the teacher supplies a term and the students illustrate and explain), but the focus should generally be developmental in nature. That is, there should be a strong emphasis on meaning and comprehension. Similarly, skills need to be practiced with regularity in order that a high level of proficiency be maintained. The focus should be developmental in nature; comprehension again is an important component.

Problem

When discussing children's performance in mathematics, frequently the comment is made that many have not mastered the basic skills. From this it is concluded that teachers do not spend enough time teaching basic computation. But this conclusion often is not valid because the inability to perform may not be associated with the initial learning but rather with a lack of maintenance. Newly learned material is particularly susceptible to being forgotten, but even material thought to be "mastered" is sometimes lost. For example, many fourth-grade teachers have had the experience in which a student seems to have mastered his basic multiplication facts; indeed, he or she can recall them with almost 100 percent accuracy; but four weeks later seems to have forgotten a great number of them.

Teaching Request

To minimize this problem and similar problems, we are asking that you incorporate review/maintenance sessions regularly into your mathematics instruction. Regularly in the sense that each Monday you have a weekly review/maintenance session and every fourth Monday you have a cumulative review/maintenance session. The purpose of the two types of review sessions is to help students retain concepts and insights.

Weekly Review/Maintenance. The following things are necessary to do if the review/maintenance component is to be implemented effectively:

1. The first half of each Monday's math period (roughly 25 minutes) should be devoted to review/maintenance.
2. The focus should be on the important skills and concepts covered in math during the previous week. The suggested order for covering these skills and ideas is:
   a. Those that are thought to be mastered and can be done very quickly.
   b. Those that need additional development and practice as iden-

tified from the analysis of the Thursday homework assignment.
c. Those that need additional work (as identified during this review session).

Most of the important skills and concepts that should be reviewed can easily be identified by examining the homework assignments from the previous week. That is, these homework assignments deal with each important data or skills; thus, reviewing them will assist you in identifying important topics. It is of utmost importance that all major ideas covered during the week be reviewed. Reviewing ideas that students have "mastered" the previous week helps to guarantee that ideas will be retained. Areas in which some reteaching is definitely needed should be identified in advance by the teacher from an analysis of the Thursday homework assignment and handled during the second portion of this designated review segment.

There are many ways that this maintenance program can be successfully organized. One important attribute of any effective organizational scheme is active student involvement. In most teaching situations, it is important to avoid situations that involve only one student in checking problems because such a procedure is usually ineffective and boring to most children. This is especially true in a review situation in which students are already familiar with the problem. A scheme that we highly recommend (because it overcomes this difficulty) is one in which the teacher presents an idea or problem and then allows students to work individually at their desks until most arrive at an answer. Finally, answers are checked (children are held accountable), and someone explains or demonstrates how to arrive at the answer (in many cases by using the chalkboard at the front of the room).

Cumulative Review/Maintenance. This aspect of the review/maintenance program can best be implemented in the following way:

1. Every fourth Monday the entire math period should be devoted to a cumulative review/maintenance session.
2. This review should encompass the work of the previous four weeks and thus replace the regular Monday maintenance/review session.

This session provides an opportunity to reteach ideas that have given difficulty over the past four weeks. It will be particularly useful to those students who have difficulty acquiring skills and ideas on initial exposure.

The interest in and value of this review session can be greatly enhanced by structuring it in an interesting format such as a game, contest, or quiz show.

Postscript

On occasion, it may be desirable to reschedule a review for a day other than Monday. For example, if by not reviewing on a Monday you can complete a chapter or unit, by all means do this and simply conduct your review on Tuesday. If it becomes necessary for you to reschedule a review, please make a note of it in the log so that we are aware of it.

MENTAL COMPUTATION

Variable

Mental computation is computation that is done without the aid of pencil and paper (or minicalculator). The process is done by the most powerful computer of all, the human brain. Mental processing is often vastly different than pencil and paper calculation. For example, in pencil and paper addition situations the calculation always goes from *right* to *left*. The student asked to solve 41 + 12 on paper is going to move mechanically from right to left. However, in a mental activity (the teacher says what is 41 + 12) the student may frequently move from *left* to *right*. First, the student does something to the tens column, then to the ones column, and then combines. We feel that the inclusion of some time for mental computation each day will help students to further develop their quantitative sense, to become more flexible in thinking and in approaching problem-solving situations. Furthermore, such activities help students to detect absurd answers (e.g., when checking their written computation) and make estimations that are frequently needed in daily activities.

Problem

The attention given to mental computation and mental problem solving has largely disappeared from the modern mathematics curriculum. At one point in time much emphasis was given to mental problem solving. This deemphasis has occurred despite some research evidence which suggests that mental practice on a regular basis appears to be related to large increases in student achievement. If students are not given some work in mental computation, then they are missing a very important way to check their work (other than the time-consuming and inefficient process of completely redoing the work).

Teaching Request

We would like you to include three to five minutes on mental computa-tion activities each day at the beginning of the lesson; the predevelopment part of the lesson will be described later in the handbook. Ideally, the material presented for mental resolution would be related to the content of the material being studied. During the study of subtraction, mental computation activities should focus on subtraction. However, some units that you study in the year will not lend themselves to this form of mental processing. During such a unit (e.g., geometry) it would be useful to rotate on a daily basis with the following types of mental computation activities: addition, subtraction, multiplication, division, and verbal problems.

The following examples will give you some ideas about the kinds of problems you may present to your students. Some of the examples here may be too easy or too difficult for your students. You should try to use problems which are challenging yet accessible to most students. It is a good idea to discuss how a problem might be solved mentally before students are asked to give solutions.

For example, for a problem like 6×12 you might suggest thinking as follows: "6 times 12, that's 6 times 10 plus 6×2, that's $60 + 12$, 72." Then begin giving students problems one at a time to solve like 8×12, 6×15, and so on. It is worthwhile to mention to the students that there are many ways to solve problems mentally and the way you showed is but one way. Children should be encouraged to discuss their mental computation procedures.

Further illustrations of the kinds of problems which are appropriate are given below. You should generate other types of mental computation exercises for your students as well.

Addition

1. $75 + 77 = $ _____

   think: $77 = 70 + 7$. First add 70 to 75 (145) then add 7 to that sum (152).

   or: Rename 77 as $70 + 7$ and 75 as $70 + 5$. Add the tens (140), add the ones (12), then find the total of the sums (152).

2. $97 + 8 = $ _____

   think: How much do I add to 97 to get 100? The answer is 3. Since 8 $= 3 + 5$, first I add 3 to 97, and then add 5 to the sum.

3. 243 + 104 = _____

   think: 104 = 100 + 4. First add 100 to 243 and then add 4 to the sum.

4. 125 + 49 = _____

   think: 49 is 1 less than 50. Since 125 + 50 = 175, 125 + 49 = 174.

Subtraction

1. 125 – 61 = _____

   think: 61 = 60 + 1. First subtract 60 from 125, and then subtract 1 from the difference.

2. 105 – 8 = _____

   think: First subtract enough from 105 to get 100: 105 – 5 = 100. Since 8 = 5 + 3, subtract more: 100 – 3 = 97.

3. 425 – 97 = _____

   think: 97 = 100 – 3. First subtract 100 from 425, and then add 3 to the difference. 425 subtract 100 is 325, add 3 is 328.

Multiplication

1. 20 × 36 = _____

   think: 20 = 2 × 10. Ten times 36 is 360, and 2 × 360 = 720.

   or: 2 × 36 = 72, so 20 × 36 = 720

   or: 20 × 36 that's the same as ($\frac{1}{2}$ × 20) × (2 × 36), or 10 × 72 = 720.

2. 4 × 17 × 25 = _____

   think: Since the product of 4 and 25 is 100, these numbers are multiplied first. Then 100 is multiplied times 17.

3. 32 × 50 = _____

   think: The product is unchanged if I double one factor and half the other factor. Thus, 32 × 50 is the same as 64 × 25 or 1,600.

4. 4 × 53 _____

   think: 53 = 50 + 3. Four times 50 is 200. Four times 3 is 12. So to find 4 × 53 add 200 + 12.

Division

1. 84 ÷ 4 = _____

    think: 84 = 80 + 4. 80 divided by 4 is 20 and 4 ÷ 4 is 1, so 84 ÷ 4 is 20 + 1 or 21.

2. 396 ÷ 4 = _____

    think: 396 = 400 – 4. Since 400 ÷ 4 = 100 and 4 ÷ 4 is 1, the quotient is 100 – 1 or 99.

3. 250 ÷ 50 = _____

    think: 250 ÷ 50 is the same as 500 ÷ 100 which is 5.

Verbal Problems

1. Mr. Thomas has a debt of $120. If he pays $70 of it, how large a debt will he have left?

    think: I need to find 120 – 70 = _____.
    12 – 7 = 5, so 120 – 70 = 50.
    50 is the answer.

INSTRUCTIONAL PACE

Variable Description

Instructional pace refers to rate. It may be thought of in terms of how quickly a class is moved through a given curriculum or in terms of how rapidly students are presented with particular topics. The pace associated with different teachers varies. Some teachers move through the curriculum faster than others.

Problem

Instructional pace may inhibit learning in several ways. At one extreme is the situation in which a teacher moves through the curriculum too quickly for learning to take place. At the other extreme is the teacher who plods along so slowly that many of the students are bored. Furthermore, some teachers, because of their slow pace, find themselves forced to cover so much material at the end of the year that they do not have time to build in the distributed practice which is essential if students are to retain the material. Research suggests that for most teachers efficiency could be improved if they increased their pace slightly. That is, there seems to be more of a tendency to procrastinate than to move forward. If the suggestions presented earlier in the manual are implemented in your teaching program, the important element of review and distributed practice should be fulfilled and you will probably be able to pick up the pace.

Teaching Request

For this variable we ask that you carefully consider your teaching behavior with respect to the instructional pace you set. Many of you will find that you can increase the pace somewhat and we ask you to attempt to do so.

The instructional strategies suggested in this study are such that if you speed up a bit too much, then you can resolve problems that arise through your regularly scheduled review/maintenance sessions.

STARTING AND ENDING THE LESSON

We have now discussed the major parts of the mathematics instructional program. Two aspects that we have not discussed explicitly are the start and end of the lesson.

The beginning portion of the lesson (*predevelopment*) will have three parts: (1) a brief review, (2) the checking of homework, and (3) some mental computation exercises. We ask that all three of these activities be done within the first eight minutes of the class period. This may be difficult for teachers who slowly ease into the lesson, but it has been commonly observed that time is frequently used inefficiently at the beginning of a lesson.

The review of the previous day's lesson should begin with a brief summary by the teacher. Several sentences that briefly and concisely remind students of what they did and why, and demonstrating how to solve a single problem are usually sufficient. Next comes the checking of homework. This should proceed very quickly once students learn that when math period begins they are to have their homework on top of their desks ready for checking. Initially, it may take some time to establish this routine, but once the routine is established it should take only a couple of minutes to check homework.

The third activity, mental computation, plays two roles in the lesson structure. First, it is an important activity per se (see earlier section). Second, these activities can provide a smooth transition for getting students engaged in thinking about math prior to the point at which the teacher begins a new development lesson.

The ending of the lesson is a very simple procedure. After allowing students a period of time for uninterrupted practice, the teacher briefly checks pupils' work on a few problems (may call on students, ask students who got problems correct to raise their hands, etc.). This accountability procedure encourages students to apply themselves during seatwork and allows an additional opportunity to clear up misunderstanding. After checking some of the seatwork, the teacher ends the mathematics lesson by assigning the homework problems.

The predevelopment phase of the lesson should take roughly eight minutes. The exact distribution of time on *review, homework*, and *mental computations* depends upon a variety of conditions (e.g., moderate difficulty with homework vs. no difficulty) and you are asked to use your judgment. In general, we think the following situation will be most applicable: one to two minutes on review; three to four minutes checking homework; and three to four minutes on mental computations.

SUMMARY AND INTEGRATION

We have asked you to do several things during the next few weeks in an attempt to improve student performance in mathematics. In the first part of this handbook we emphasized that we didn't feel that changing one or two teacher behaviors would make much difference in student performance. We feel that the systematic application of all the behaviors discussed in this treatment program can make an important difference in student learning. The purpose of this last section is to briefly review the teaching requests we have made and to show how the pieces fit together into a total program.

The predevelopment portion of the lesson begins with a brief summary and a review of the previous lesson. The review (including the checking of homework) is designed to help students maintain conceptual and skill proficiency with material that has already been presented to them. Mental computation activities follow and provide an interesting bridge for moving into the new lesson.

Next comes the development part of the lesson, which is designed to help students understand the new material. Active teaching helps the student comprehend what he or she is learning. Too often students work on problems without a clear understanding of what they are doing and the reasons for doing it. Under such conditions, learning for most students will be filled with errors, frustration, and poor retention. If student learning is to be optimal, students must have a clear picture of what they are learning; the development phase of the lesson is designed to accomplish this understanding.

The controlled practice that occurs toward the end of the development portion of the lesson is designed to see if students are ready to begin seatwork. It simply doesn't make sense to assign seatwork to students when they are not ready for it—practicing errors and a frustrating experience guarantee that student interest and performance in mathematics will decline. The controlled practice part of the lesson provides a decision point for the teacher. If students generally understand the process and are able to work problems correctly, then the teacher can proceed to the seatwork portion of the lesson. If student performance on problems is relatively poor, then the development must be retaught. If students are ready to do practice work, it is foolish to delay them; similarly, if students are not ready to do development work, it is foolish to push them into it. The controlled practice part of the lesson allows the teacher to decide if it is more profitable to move to seatwork or to reteach the development portion of the lesson.

Hence, when teachers move to the seatwork portion of the lesson, students should be ready to work on their own and practice should be relatively error free. Seatwork provides an opportunity for students to practice successfully the ideas and concepts presented to them during the development portion of the lesson. If teachers consistently present an active development lesson and carefully monitor student performance during the controlled portion of the lesson, then student seatwork will be a profitable exercise in successful practice.

The seatwork part of the lesson allows students to organize their own understanding of concepts (depend less upon the teacher) and to practice skills without interruption. The seatwork part of the lesson also allows the teacher to deal with those students who have some difficulty and to correct their problems before they attempt to do homework. If teachers actively monitor student behavior when seatwork is assigned and if they quickly engage them in task behavior and maintain that involvement with appropriate accountability and alerting techniques, then the essential conditions have been created for successful practice.

Homework is a logical extension of the sequence we have discussed. During the mathematics lesson students learn in a meaningful setting. During seatwork students have a chance to practice and deal with material they understand. The homework assignment provides additional practice opportunity to further skill development and understanding.

The above aspects of the mathematics lesson combine to give the students: (1) a clear understanding of what they are learning; (2) controlled practice and reteaching as necessary to reinforce the original concepts and skills; (3) seatwork practice to increase accuracy and speed; and (4) homework assignments which allow successful practice on mastered material (distributed practice, which is essential to retention).

To maintain skills it is important to build in some review. Skills not practiced and conceptual insights not reviewed from time to time tend to disappear. Even mature adults forget material and forget it rapidly. For this reason we have asked you to provide for review of material presented the previous week each Monday and to provide a comprehensive review

TABLE 3.2　Weekly Lesson Time Table

Monday	Tuesday	Wednesday	Thursday	Friday
Weekly Review (20 Min.)	Homework, Review, Mental Computation (8 Min.)	Homework, Review, Mental Computation (8 Min.)	Homework, Review, Mental Computation (8 Min.)	Homework, Review, Mental Computation (& Min.)
Development (10 Min.)	Developmental (20 min.)	Developmental (20 min.)	Developmental (20 min.)	Developmental (20 min.)
Seat Work (10 Min.)	Seat Work (15 min.)	Seat Work (15 min.)	Seat Work (15 min.)	Seat Work (15 min.)
Lesson Conclusion & Homework Assign. (2 Min. Max)	Lesson Conclusion & Homework Assign. (2 Min. Max)	Lesson Conclusion & Homework Assign. (2 Min. Max)	Lesson Conclusion & Homework Assign. (2 Min. Max)	Lesson Conclusion (2 Min. Max.)

TABLE 3.3 Lesson Time Table (4th Week)

Monday	Tuesday	Wednesday	Thursday	Friday
Monthly Review (45 min.)	Review, Mental Computation (8 Min.)	Homework, Review, Mental Computation (8 Min.)	Homework, Review, Mental Computation (8 Min.)	Homework, Review, Mental Computation (8 Min.)
	Developmental (20 min.)	Developmental (20 min.)	Developmental (20 min.)	Developmental (20 min.)
	Seat Work (15 min.)	Seat Work (15 min.)	Seat Work (15 min.)	Seat Work (15 min.)
	Lesson Conclusion & Homework Assign. (2 Min. Max)	Lesson Conclusion & Homework Assign. (2 Min. Max)	Lesson Conclusion & Homework Assign. (2 Min. Max)	Lesson Conclusion (2 Min. Max.)

every fourth Monday. Such procedures will help students to consolidate and retain their learning. Finally, we have suggested that the systematic presentation of mathematics material may facilitate student learning (i.e., initial acquisition) such that you can pick up the pace a bit and we encourage you to do so if you can. Finally, when many students experience trouble, the development portion of the lesson should be repeated and students should never be asked to do homework until they are ready to do it successfully.

The plan described above is summarized in Table 3.2. This table outlines the sequence and length of each lesson component in order to provide a general picture of the mathematics lesson that we are asking you to teach. Table 3.3 shows the pattern of instruction for a week containing a comprehensive review. This pattern occurs every fourth week.

APPENDIX 3.A

Process/Product Questions

VARIABLE DESCRIPTION

Process questions ask the student to explain something in a way that requires him or her to integrate facts or to show knowledge of interrelationships. Process questions often begin with *why* or *how* and can't be answered with one word. Many process questions require the student to specify the cognitive and/or behavioral steps that must be gone through in order to solve a problem or come up with an answer. Two examples of process questions follow. "Allen, if a man bought 3 tickets for $2.85 and 2 tickets for $2.15 and if we wanted to know the average cost per ticket, how could we get the answer?" Similarly, if the teacher asks, "There are 60 minutes in an hour, how can we find out how many minutes in ¼ hour?", she or he is asking a process question. The student is asked to explain a process and to verbalize understanding ("We can always find ¼ of anything by dividing by 4.").

Product questions only require a knowledge of a specific fact and can often be answered with a single word or by providing a number (answer to a problem). Product questions often begin with the words *who, what, when, where, how much, how many,* etc. A written example of a product question would be: $7 + 3 = $ _____? An oral product question would be: "Zero times seven equals how much?"

Product questions can be transformed into process questions by asking for an explanation rather than an answer: "Why does zero times seven equal zero?" The child is being asked to show awareness of the principle by saying something like, "When zero is a factor the product is zero"; or "Zero times anything equals zero." A written example of a process question would be: "$7 + 3 = 10$ and $3 + 7 = 10$; why?" The student is expected to respond with something like, "Changing the position (order) of the addends (numbers) does not change the sum."

In summary, product questions are those questions that ask students to provide the right answer (*how much, what, when*). In contrast, process questions ask students to explain how an answer was or could be obtained (*why* questions).

PROBLEM

Often when teachers think about development and conceptual work, they equate it with process questions. This is *not* the case. Indeed, often process questions are overused or used inappropriately. The problem with process questions is that they are sometimes ambiguous to the student (What is the teacher asking me?) and may produce an ambiguous student response even though the student understands the concept. Process questions often consume a lot of instructional time (student thinks, mentally practices the response, makes an oral response). Hence, if process questions are overused, a lot of instructional time can be wasted. If selectively used, process questions can be very valuable. For example, by asking a few process questions, teachers can see if students understand the rationale or principle upon which computational work is based and help consolidate student learning.

If teachers are alert to student responses, hold students accountable by asking individual students questions, and keep all students involved in the lesson, then the learning of unproductive habits is minimized. If process and product questions are used appropriately, then student involvement and achievement are enhanced. If they are used inappropriately, then much instructional time is lost and errors are practiced—errors that subsequently are very hard to correct.

REQUEST FOR TEACHING BEHAVIOR

We feel that the presence of a few process questions in the development stage of a lesson is helpful (especially when a new principle is being introduced) because listening to a students' explanations can help teachers diagnose inappropriate assumptions, etc., that students have made. However, we believe that most of the process development can be done through teacher modeling of process explanations rather than by asking students to respond to process questions. For example, the teacher could ask, "Who can tell me what zero times seven is?" The teacher surveys the room and calls upon Bill (who may or may not have his hand up). When Bill says, "Zero," the teacher could respond with something like, "That's right, Bill, the answer is zero. Whenever zero is a factor, the product is always zero." By actively verbalizing and demonstrating (e.g., writing problem solutions on the board, etc.), teachers can help students to achieve process understandings in a very efficient way. Still, it is useful to ask process questions occasionally to assess student understanding. However, if asked properly, product questions can provide information that assesses the student's ability to relate ideas, transfer concepts to different situations, and understand the process sufficiently well to solve problems. Product questions can also provide all students in the class (or group) a chance to practice the computation. This is especially true when the teacher asks the question first and then calls on a student. If a teacher names a student and then asks the question, many of the students will not perform the calculation (that's Mary's problem). Similarly, if teachers hold nonvolunteers accountable on occasion, it increases the number of students who are likely to think about the problem under discussion.

Although a major goal of the development portion of the lesson is to strengthen students' conceptual understanding (*why*), this goal can be achieved with a heavy use of product questions. The usefulness of product questions is due to the following factors: (1) They typically elicit a quick response from the student (and quick feedback from the teacher); hence, more material can be covered in a given amount of time. (2) They provide more practice opportunity for a broader number of students; hence, a teacher's diagnosis is not limited to the responses of a few students. (3) They help to create a "can do" attitude on the part of students (a series of quick questions that the students respond to successfully). However, it is desirable to ask process questions and enter a diagnostic cycle (reteaching) when students respond to product questions incorrectly. When students miss the same type of product questions, then it is useful to stop and review the process and ideas behind the computation. To reiterate, process questions can and do play a valuable role in successful mathematics teaching although they should not be overused.

Typical Homework Assignment

Reproduced below is a page from the teachers edition fourth-grade Holt mathematics textbook. An appropriate homework assignment would be to assign problems #4 through #18 (evens). The remaining problems could be used in connection with the development or seatwork portions of the lesson. Appendix 3.E shows how these same problems could be put in a different format and thus provide some variety in your assignments.

Exercises

Add. Look for patterns.

1.	3	13	23	43	73
	+ 6	+ 6	+ 6	+ 6	+ 6
	9	19	29	49	79

2.	4	14	24	64	84
	+ 7	+ 7	+ 7	+ 7	+ 7
	11	21	31	71	91

Add.

3.	41	4.	65	5.	93	6.	14
	+ 2		+ 2		+ 6		+ 5
	43		67		99		19

7.	23	8.	41	9.	65	10.	84
	+ 8		+ 9		+ 6		+ 9
	31		50		71		93

11.	84	12.	36	13.	48	14.	36
	+ 6		+ 9		+ 8		+ 7
	90		45		56		43

Solve these problems.

15. 17 cents for candy.
8 cents for gum.
How much in all?
25 cents

16. 76 players.
3 more joined.
How many now?
79 players

17. 35 pounds of oranges.
9 pounds of apples.
How much fruit?
44 pounds

18. 24 bees.
8 ants.
How many insects?
32 insects

Variety in Assignments

Frequently students can be freed from the somewhat boring routine of always doing problems from the textbook as their homework assignment. The assignment shown below is an alternate to the typical row-by-row set of computation exercises found in most textbooks, yet it accomplishes the same objectives in a more interesting format. Answers for the problems are shown in parentheses.

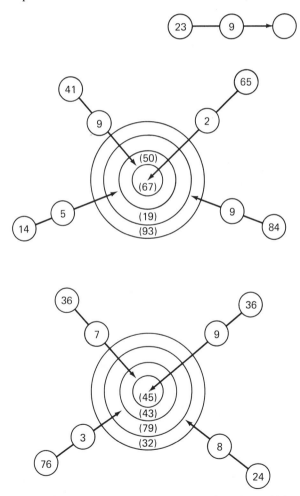

Add to find the missing target values. For example, 32 would be the missing value in this example:

APPENDIX 3.D

Systematic Processing Errors Illustrations

A systematic processing error is an error a student consistently makes on a particular kind of problem. It is different from making random errors. Simple examples include always working addition from left to right or "borrowing" in every subtraction problem whether or not it is necessary. Other common examples are explained below.

In each of the following situations, carefully analyze the examples and try to determine the error pattern. Then check your work by reading the description of the error pattern.

Situation #1

23	34	29	38
+ 6	+ 9	+ 5	+ 4
83	124	79	78

Error Pattern: In these problems the student does not add straight down a column, but rather adds the number of tens from the first number to the units from the second number. Thus, in example #1 the 2 tens are added to the 6 ones to get 8 tens.

Situation #2

53	86	95	31
− 27	− 39	− 27	− 19
34	53	72	28

Error Pattern: In these problems the student does not "borrow," but rather always subtracts the smaller digit from the larger digit.

Situation #3

7	5	3	5
48	49	86	67
× 59	× 36	× 45	× 28
432	294	430	536
270	177	354	174
3132	2064	3970	2276

62

Error Pattern: The first part of each problem, the multiplying by the ones, is done correctly. However, when multiplying by the tens the crutch number recorded from the multiplying by ones is incorrectly used again. For instance, in the first example, when multiplying by the 5 tens the 7 (carried over from the 9 × 8) is used again when the 7 is added to 5 × 4 and the 27 is recorded.

Situation #4

4	1	3	2
26	83	38	53
× 7	× 5	× 4	× 8
422	455	242	564

Error Pattern: In these problems the crutch is added before multiplying in the tens place, whereas the correct procedure is to multiply and then add the crutch. Thus, in the first example the 4 is added to the 2 and then this sum multiplied by 7. If this problem was done correctly, the 2 is multiplied by the 7 and then the 4 is added.

Situation #5

44	14	87	39
2⟌88	4⟌164	3⟌234	5⟌465
80	160	210	450
8	4	24	15
8	4	24	15

Error Pattern: These problems are worked correctly except that the quotient figures are written from right to left. Consider the third example, there are 7 threes in 23, but the 7 is recorded at the extreme right, rather than above the 3.

Situation #6

32r3	78r2	94r2
9⟌2721	6⟌4250	6⟌5426
27	42	54
21	50	26
18	48	24
3	2	2

Error Pattern: In these problems, whenever the student brings down and cannot divide, he brings down again but forgets to record a zero in the quotient.

Letter to Parents

August 25, 1977

Dear Parent(s):

As part of the fourth grade math instructional program this year, I will be regularly assigning some work for the students to complete at home. It should take your son or daughter about fifteen minutes to complete this homework. If you find that it regularly takes considerably longer for him/her to finish this assignment or the assignment causes other difficulties, please let me know in that I may be assigning too many or too difficult problems.

Programs in other school districts, educational research, and common sense indicate that the more a student practices important math concepts and problems, the more proficient he/she becomes in essential math skills. I view homework as an opportunity for the student to practice the concepts and skills that he/she has learned in class. I hope that you will encourage your son or daughter to complete every assignment to the best of his/her ability. Parental support is very helpful. Thank you for your cooperation in this matter.

Sincerely,

Teaching Groups in Schools Using a Department Organization

The emphasis thus far has been placed upon teaching mathematics to the class as a unit. We feel that many of the principles presented (the importance of development, the use of controlled practice and seatwork, accountability, etc.) will transfer to classrooms in which teachers are teaching groups of students. In applying these principles to a group situation, teachers will have to adjust them to their teaching situation.

In general, we are not enthusiastic about the use of two or more groups to teach mathematics. Three recent and major research projects have shown that third-, fourth-, and fifth-grade students appear to benefit more from whole-class instruction than they do from individual or group instruction. Although the precise reasons for these differences are unknown, we suspect that students learn less in group and individual settings because they have less direct developmental work with the teacher. Also, the extra transitions (teachers moving from group to group) probably result in the loss of time that could have been used for instructional purposes. Furthermore, student work is probably less effective when the teacher is not available to supervise work.

If the differences between groups are not great, we strongly recommend that the class be taught as a whole class. However, we understand that sometimes the differences between students in a given classroom are so great that grouping is a practical necessity.

If grouping is necessary, you should attempt to limit yourself to only two groups because the transition and supervision problems that accompany the use of more than two groups are normally very difficult to justify.

Since teaching circumstances are so varied (sometimes the difference between two groups is moderate but in other classrooms there are vast differences between the two groups), it is impossible for us to describe a plan that would be best in all situations. Still, there are a few key things that we would like to emphasize.

First, whenever possible, we think it will be useful for you to teach the class as a group. Students learn a great deal from teacher illustrations and explanations. Perhaps the easiest way to do this in a group situation is by holding common reviews from time to time. The review might be a short-term review for the lowest group and a long-term review for the highest group.

An especially good way to conduct a common review is through the use of mental computation problems. We strongly recommend that each day of the week but Monday you use the first ten minutes of the class for review with mental computation problems. As we have noted earlier in the handbook, we feel that mental computation problems are a very important addition to an instructional program.

Second, we would like you to set aside each Monday for a review session. After spending the first five minutes on mental computation, review ideas and skills that are needed by both groups. Then involve one group in a seatwork review, and begin the developmental review with the other group. Roughly half way through the period reverse the roles; give group two a seatwork review assignment and begin an oral review with group one.

To maximize the value of this review, a homework assignment containing review problems should be given the previous Thursday. Your analysis of these papers should suggest the topics and skills that should receive emphasis in the Monday review. Besides the homework assignment each Thursday, we request that you assign homework three other days per week. Remember that these assignments are to provide brief, successful practice.

The third request is that you maximize the amount of development time for each group. The exact amount to be given to each group will necessarily vary depending on the topic being considered and the group itself; however, the importance of development work for both groups cannot be overemphasized. As you do the development work, remember the guidelines previously discussed. For instance, teacher explanations and illustrations are important, especially initially. Also, process explanations are very important and oftentimes are related to efficient use of limited instructional time.

Finally, we ask that you implement other recommendations as regularly and consistently as you can. Little things are important (e.g., getting all students started on seatwork before doing other instructional tasks) and we hope you will carefully review the ideas presented in the handbook with an eye toward applying them in your classroom.

4

Elementary School Experiment I

The field testing of the instructional program commenced in the fall of 1977. With the active assistance of administrators and principals in a large public school system, it was possible to recruit a volunteer sample of 40 fourth-grade teachers who used the semidepartmental plan (teachers teach only two or three different subjects a day). The decision was made to do the research within this organizational pattern because it afforded a classroom structure that was most comparable with the classroom organization in which the correlational research was conducted (e.g., no classrooms were completely individualized). Choice of this structure also provided a rough control for instructional time, since teachers did not keep their mathematics students for the entire day. Hence, for most of the teachers there was pressure to end the mathematics class at a set time, and reteaching later in the day was impossible.

In the 1975 naturalistic study, emphasis was placed upon internal validity. We chose a relatively stable school district (few changes in teacher personnel) in which a common textbook was used in all classrooms, and where student populations were comparable across schools and classrooms. The reason for these controls was to exclude as many rival hypotheses as possible to the conclusion that teachers and teaching were affecting student learning. In the 1977 experimental study, external validity was emphasized. A more heterogeneous school population was sampled because we felt this would be a more legitimate test of the training program.

Teacher Training

On September 20, we met with all teachers and all school principals who had volunteered to participate in the project as well as the associate superintendent for instruction and many of the supervisors in the school district. Eighty percent of the teachers available volunteered for the program. Many of the semidepartmentalized schools were in low socioeconomic status (SES) areas.

At the initial workshop, all 40 teachers were told that the program was largely based upon an earlier observation of relatively effective and ineffective fourth-grade mathematics teachers. Teachers were told that although we expected the program to work, the earlier research was correlational and the present project was a test of those ideas. After a brief introduction, the teachers (drawn from 27 schools) and their principals were divided into two groups: treatment and control. (Schools were used as the unit for random assignment to experimental conditions.) This was done to eliminate the difficulties that would doubtlessly follow by implementing the treatment in one class but not another in the same school.

Teachers in the treatment group were given an explanation of the program (the training lasted for 90 minutes). After the training session, they were given the treatment manual (see previous chapter), along with instructions to read it and to begin to plan for implementation. Two weeks after the program began we returned to meet with treatment teachers. The purpose of this 90-minute follow-up meeting was to respond to questions that teachers had about the meaning of certain teaching behaviors and to react to any difficulties that the teachers might have encountered. Thus, the treatment consisted of two 90-minute training sessions and a program manual that detailed the treatment and provided a base for teacher reference as necessary.

Control teachers were told that they would not get the details of the instructional program until February, 1978. Furthermore, they were told that it was hoped that this information might be especially useful to them then because at that time they would receive individual information about their own classroom behavior and refined information about the program itself. Finally, control teachers were told that their immediate role in the project was to continue to instruct in their own style.

Because control teachers knew that the research was designed to improve student achievement, that the school district was interested in the research, and that they were being observed, we feel reasonably confident that differences in performance between control and treatment groups were due to the program, and not simply increased motivation of the experimental teachers. Sometimes experiments appear to work but in actuality they work only because of motivational or Hawthorne factors (participants try harder because they know that they are in an experiment) and we took steps to eliminate this possibility. However, at the other extreme, we did not intend to create so strong a "press" that control teachers would seek out information from treatment teachers or would alter their instructional style in an effort to guess what the experimenters wanted. If control teachers changed their instructional behavior, then differences could be due to the fact that previously they had been using a poorly thought out or inconsistent pattern of instruction.

Process and Product Measures

The treatment program started on October 3, 1977, and was terminated on January 25, 1978. During the course of the project all 40 teachers (with few exceptions) were observed on six occasions. Observers collected information using both low- and high-inference process measures. The basic nature of the observational instruments was described briefly earlier in the discussion of the background study. For more details see Good and Grouws (1975a). Reliabilities of all observational variables in the present study were at least .80. Students were administered the mathematics subtest of a standardized achievement test (Science Research Associates; SRA, Short Form E, blue level: KR 21 Reliability Estimate .80) in late September and in mid-December. In mid-January students responded to a mathematics achievement test constructed by Robert E. Reys at the University of Missouri (KR 21 Reliability Estimate .78). This test measured the content that students had been exposed to during the program. Furthermore, an instrument measuring student learning styles (preferences and attitude toward mathematics) was administered in September and in January (see Appendix A). The development of this instrument and the reliability of scales will be discussed later. Also in January, a ten-item attitude scale was administered (reliability .89) to assess the impact of the treatment on students' attitudes toward mathematics (see Appendix A, last 10 questions of part 1). An instrument measuring teacher beliefs and preferences for mathematics instruction was also used in September. (See Appendix A. The development of this instrument and its reliability will be discussed later; see Chapter 7.)

RESULTS OF EXPERIMENTAL STUDY I

At the debriefing session in February, control teachers consistently indicated that (1) they did think more about mathematics instruction this year than previously, (2) they did not feel that they had altered their behavior in any major way, and (3) directly or indirectly they had not been exposed to program information. Hence, adequate control conditions appeared to have been satisfactorily implemented.

Implementation

If one is to argue that a program works or is responsible for a change, it is important to show that treatment teachers exhibited many more of the classroom behaviors related to the treatment than did control teachers. In this study treatment teachers were found to perform significantly more of the treatment behaviors than did control teachers. To assess the degree of implementation of the experimental treatment, coders collected low-inference data concerning the presence, absence, or duration of specific instructional events in control and experimental classrooms. Results from reliability estimates revealed 90 percent or better agreement on each of the variables employed to assess treatment implementation.

Given the complexity (several different behavioral requests involving sequences of behaviors) of the treatment, it is difficult to provide a single, precise statement about the extent to which the treatment was implemented. Implementation was estimated by using a summary checklist that observers filled out at the end of each observation. The information on the checklist provides good but not total coverage of treatment behaviors.

To illustrate the scoring procedure more concretely, one of the eight scoring definitions will be explained in detail. The following variables are relevant to that scoring procedure:

Daily Review: The number of minutes the teacher devoted toward a review of the previous day's assignments, developed concepts of skills, etc. Coded as number of minutes.

Development: The number of minutes the teacher devoted to establishing comprehension of skills and concepts in a direct manner. Coded as number of minutes.

Seatwork: The number of minutes the students practiced work individually or at their desks. Coded as number of minutes.

Review: The number of minutes the teacher devoted toward review of any type. Coded as number of minutes.

Involvement during Seatwork: The number of students clearly on task divided by the total number of students.

Availability of Teacher during Seatwork: The availability of the teacher to help students with their seatwork and active demonstration on the teacher's part of a willingness to be approached for help. Coded as yes or no.

Student Accountability: Actual checking of students' seatwork toward the end of the seatwork phase of the lesson. Coded as yes or no.

Homework: The assignment of homework. Coded as yes or no.

Mental Computation: The teacher asking the students to work in their heads; a problem given verbally or written on the board. Coded as yes or no.

To obtain overall score for implementation for each teacher, scores for each variable were first averaged across observations, then all scores were combined. Since the treatment program actually specified time parameters for daily review, development, seatwork, and total review, an adjustment was made to allow for these program specifications. Scores on these four variables represent actual minutes spent on these activities up to a specified limit. Time in excess of this limit was not credited to the score. (Limits were daily review = 8 minutes, development = 20 minutes.) To adjust for the scaling differences among these four scales and those that were *dichotomous* (yes or no), each score was divided by the time limit for that variable, thus reducing the range to a minimum of zero and a maximum of one. This corresponded to the scale ranges of the remaining dichotomous variables. Finally, because the time variables were felt to be a more important indicator of implementation than the dichotomous variables, the scores were multiplied by two. An arithmetic representation of this process is presented below.

$$\text{Implementation Score} = \frac{2(\text{Daily Review})}{8} +$$

$$\frac{2(\text{Development})}{20} + \frac{2(\text{Seatwork})}{16} + \frac{2(\text{Total Review})}{12} +$$

+ Involvement during Seatwork + Availability of Teacher during

Seatwork + Student Accountability + Homework + Mental

Computations

The particular method described above for estimating program implementation is the most conservative method for estimating program use. The implementation scores for the experimental and control teachers as well as the means and standard deviations (using this conservative formula for estimating program implementation) are presented in Table 4.1. Results of the analysis of variance test comparing control and experimental teachers also are presented in Table 4.1. The results of this test indicate that the experimental teachers did exhibit more of the behaviors called for in the Missouri Mathematics Program than did control teachers. Comparisons of implementation between treatment and control teachers using

TABLE 4.1 Analysis of Variance between Experimental Treatment and Control Treatment Teachers' Implementation Scores*

Source	df	MS	F	Probability
Treatment Condition	1	13.28	4.53	0.0400
Error	37	2.93		

*The mean for the control group was 7.89 (S.D. 2.02) and for the treatment group 9.06 (S.D. 1.35).

other scoring definitions consistently illustrate that across all definitions treatment teachers were found to use more of the behaviors called for in the treatment program than did control teachers.

Table 4.2 summarizes some of those behaviors included on the checklist which were used to estimate the degree of program implementation. For example, in 91 percent of the observations, treatment teachers were found to conduct a review, whereas control teachers were found to conduct a review 82 percent of the time. There were only 2 of the 21 treatment teachers who exhibited uniformly low implementation scores. (Data from these two classrooms were left in the analysis to assume the most conservative procedure for estimating treatment effects.) Table 4.2 also reports the correlation between the frequency of occurrence of selected treatment behaviors and teacher residual gain scores on the SRA mathematics test. As can be seen, homework assignments, frequent review, and use of mental computation activity were found to correspond with favorable gains.

Development appears to be the only variable that teachers had consistent trouble implementing. The reason for the low level of implementation may be that teachers focused on the many other teaching requests that were perhaps easier to implement. Alternatively, teachers might not have had the knowledge base necessary to focus on development for relatively long periods of time. Another possibility is that some of the other components required more time and preparation than we anticipated and thus development was given insufficient attention by the teachers. These possible explanations need further study, but the results of studies in which development alone was manipulated clearly suggest that a development component is important. More work needs to be done on this concept, and it may involve different types of teacher training programs.

Impact on Student Performance

As can be seen in Table 4.3, the treatment group began the project with lower achievement scores than did the control group. The initial difference between experimental and control students was significant ($p < .001$). These figures show that after the 2½ months of the project, the number of questions that was answered correctly by the average student in the experimental group increased from 11.94 to 19.95. Similarly, it can be seen that in terms of national norms, the percentile rank of the experimental group increased from a percentile of 26.57 to 57.58. Such results are truly impressive, considering the comparatively short duration of the project. Interestingly, the control group also shows a large gain, but their gains do not match those of the experimental group.

All experimental teachers taught mathematics to the class as a whole (as requested). However, only 12 of the 19 control teachers taught

TABLE 4.2   Mean Percent of Occurrence of Selected Implementation Variables for Treatment and Control Group Teachers and the Correlation of These Variables with Teachers' Residualized Gain Scores on the SRA Mathematics Test

	Treatment $\bar{x}$ (percent)	Control $\bar{x}$ (percent)	p-Value	Correlation	p-Value
1. Did the teacher conduct review?	91%	62%	.0097	.37	.04
2. Did development take place within review?	51%	37%	.16	.10	.57
3. Did the teacher check homework?	79%	20%	.0001	.54	.001
4. Did the teacher work on mental computation?	69%	6%	.001	.48	.005
5. Did the teacher summarize previous day's materials?	28%	25%	.69	.20	.26
6. There was a slow transition from review.	7%	4%	.52	-.02	.91
7. Did the teacher spend at least 5 minutes on development?	45%	51%	.52	-.08	.65
8. Were the students held accountable for controlled practice during the development phase?	33%	20%	.20	.12	.50
9. Did the teacher use demonstrations during presentation?	45%	46%	.87	-.15	.41
10. Did the teacher conduct seatwork?	80%	56%	.004	.27	.13
11. Did the teacher actively engage students in seatwork (first $1\frac{1}{2}$ minutes)?	71%	43%	.0031	.32	.07
12. Was the teacher available to provide immediate help to students during seatwork (next 5 minutes)?	68%	47%	.02	.28	.11
13. Were students held accountable for seatwork at the end of seatwork phase?	59%	31%	.01	.35	.05
14. Did seatwork directions take longer than 1 minute?	18%	23%	.43	-.02	.92
15. Did the teacher make homework assignments?	66%	13%	.001	.49	.004

TABLE 4.3 Preproject and Postproject Means and Standard Deviations for Experimental and Control Classes on the SRA Mathematics Achievement Test, Field Experiment I

I. All Treatment and all Control Teachers

	Preproject Data			Postproject Data			Pre-post Gain		
	Raw Score	Grade Equivalent	Percentile	Raw Score	Grade Equivalent	Percentile	Raw Score	Grade Equivalent	Percentile
Experimental:									
Means	11.94	3.34	26.57	19.95	4.55	57.58	8.01	1.21	31.01
Standard deviations	3.18	.51	13.30	4.66	0.67	18.07			
Control:									
Means	12.84	3.48	29.80	17.74	4.22	48.81	4.90	0.74	19.01
Standard deviations	3.12	.48	12.43	4.76	0.68	17.45			

II. Control–Whole-Class Teachers and Control-Group Teachers

	Preproject Data			Postproject Data			Pre-post Gain		
	Raw Score	Grade Equivalent	Percentile	Raw Score	Grade Equivalent	Percentile	Raw Score	Grade Equivalent	Percentile
Whole-Class Control:									
Means	11.70	3.30	25.30	16.20	3.98	43.00	4.50	0.68	17.70
Standard deviations	2.58	.40	10.15	4.96	0.68	18.09			
Group Control:									
Means	14.78	3.77	37.50	20.38	4.64	58.77	5.60	0.87	21.27
Standard deviations	3.14	.48	12.68	3.12	0.47	11.56			

NOTE: SRA = Science Research Associates.

mathematics to the whole class, whereas the other 7 taught mathematics to groups of students. Hence, pretest and posttest differences for control whole-class and control-group teachers are presented separately in Table 4.3. As can be seen, control-group teachers started and ended the project with greater student achievement levels than did control whole-class teachers and the experimental teachers. However, it should be noted that the raw achievement gain of the experimental teachers was much higher than that of the control teachers who taught groups of students. In fact, in 2½ months, the experimental classes virtually caught up with the classes of control teachers who used a group strategy.

Table 4.4 presents the results from an analysis of variance on residual gain scores comparing the performance of experimental and control groups. The performance of the treatment group significantly exceeds the performance of the control group. All of the residual means show a large positive discrepancy for the treatment group. That is, the experimental group showed considerably more achievement at the posttesting than was predicted by the pretest. In contrast, the control group showed a large

TABLE 4.4   Analysis of Variance on Residual Gain Scores (Using Mean Teacher Scores) for Treatment and Control Teachers on SRA Test and Content Test

I. SRA Mathematics Achievement Test
A. Treatment vs. Control (Group Teachers Included)

	Treatment	Control	p-Value
Grade level scores	2.22	−2.08	.002
Percentile scores	5.67	−5.51	.003
Raw scores	1.53	−1.46	.002

B. Treatment vs. Control (Group Teachers *Not* Included)

	Treatment	Control	p-Value
Grade level scores	1.98	−3.31	.002
Percentile scores	5.11	−8.46	.003
Raw scores	1.30	−2.22	.002

II. Content Mathematics Test*
A. Treatment vs. Control (Group Teachers Included)

	Treatment	Control	p-Value
Content test	1.14	− .48	.10

B. Treatment vs. Control (Group Teachers *Not* Included)

	Treatment	Control	p-Value
Content test	1.13	− .83	.11

*Analyses using the student rather than the teacher as the unit of analysis yielded the following results: Treatment vs. Control (group teachers included), $p < .008$; Treatment vs. Control (group teachers *not* included), $p < .002$.

negative discrepancy. Table 4.4 also indicates that the performance of the experimental group exceeds that of the control group (with and without group teachers included) on the residual mathematics content test scores (using SRA raw scores as the covariate).

The favorable results reported here were also confirmed by data analyses performed by the school district. In part, their testing interest was stimulated by the following two factors: (1) the test period was very short, and (2) the SRA short form was used and does not yield separate scores for concepts and computation. To address these concerns, the evaluation office of the public school system ran a covariance comparison using the long form of the SRA (using students' scores on the April, 1977 test as a covariate) to assess the April, 1978 performance of treatment and control students. The analyses on both the SRA Concepts and Computation Tests revealed a significant difference ($p<.02$) in favor of students in the treatment group. These data collected by the school district in its regular testing program demonstrated that the treatment effects continued three months after the project terminated (i.e., post testing was completed and observation stopped).

Correlations between teachers' implementation scores and residual gain performance on the standardized achievement test and the mathematics content test were computed. All of the implementation definitions correlated positively with residual gain performance; however, the correlations were consistently higher between implementation scores and performance on the standardized test than on the mathematics content test.

The lower correlations between implementation and the content test may be due to the procedures used in constructing the content test. The plan was to assemble a test that measured the content to which most students had been exposed. Some teachers appear to have gone ahead in the textbooks to material that the test did not measure. It may be that teachers who had high gains on the standardized test were penalized by a ceiling effect on the content test.

In addition to the statistical analyses presented, it is useful to consider teachers' rank order in the distribution of residual scores. For example, within the control group, teachers who used a whole-class teaching strategy obtained both the best and worst results. Three of the five control teachers who had the highest residual means taught the entire class; however, the lowest six teachers also taught with a whole-class methodology. These results are a direct replication of our earlier naturalistic research (Good and Grouws, 1975a; 1977), in which it was also found that teachers who used a group teaching strategy fell in the middle of the distribution of residual gain scores.

Examination of teachers' rank order in the distribution of residual gain scores for achievement also helps to illustrate the general effectiveness of the treatment. Ten of the 12 teachers with the highest residual means were in the treatment group, and none of the treatment teachers was among the 5 lowest teachers. However, the impact of the treatment is not even across the treatment group. Some teachers show considerably less gain relative to other teachers. Strong emphasis should be placed on the word *relative*, because all teachers' posttest means were higher than their pretest means. Also, it should be noted that a sizeable, positive correlation (.64) was found between teachers' residual scores on the SRA test and the content test. Teachers who are high (or low) on one measure tend to be high (or low) on the other. Hence, in this particular study the content coverage of the standardized achievement test appears to correspond reasonably well with the curriculum taught. (This is not always the case, see Freeman et al., 1980.)

Student Attitude and Teacher Attitude

The experimental students also expressed more favorable attitudes toward instruction than did control students on a brief attitudinal questionnaire. The mean (lower scores indicate greater satisfaction) for the control group was 18.38 and the mean for the experimental group was 17.55 ($p<.05$). The differences were statistically significant although the practical differences are small. The data suggest, at a minimum, that the achievement gains did not come at the expense of attitudes, at least within the limits of the operational attitude measure. It would seem that emphasis upon variables like review and homework (when done in the context of meaningful and successful practice) does not necessarily lower attitudes as it is sometimes argued.

It is also important to note that feedback from the experimental teachers was supportive and indicated that teachers felt the program was beneficial and that they planned to continue using it. We collected this information because we feel that teachers' feelings and beliefs about an instructional program are as important as teacher behaviors. Despite the fact that student achievement improves, teachers may choose not to continue an instructional program because it takes more time for preparation, makes teaching too demanding, or because it conflicts with teachers' personal definitions of what teaching should be. Thus, we found the anonymous positive feedback of teachers very edifying because it gave a sense of ecological validity to the research.

The letter and response form that were sent to teachers to gather program feedback information can be found in Appendix F. Twenty of the 21 experimental teachers responded to the confidential letter (they were provided with a self-addressed, stamped envelope, and no numbers, names, or identifying information was associated with the survey instrument).

Eighteen teachers felt that all six phases of the project were either "very good" or "good" (valuable to them as teachers). The regular assignment of homework proved to be the most useful methodology, while increased pace invoked the lowest affective response from teachers. Even so, 13 of 17 teachers responding to this item thought that the increased pace stage of the project was "good."

Questionnaire responses revealed that most of the participants planned to continue using all aspects of the program, on or near the initial level recommended by the project directors. After the program had ended, 18 of 20 teachers were still conducting expanded weekly review sessions. At least 14 teachers were still implementing the prescribed development and mental computation phases four to five times a week, and 15 teachers continued to assign homework at least three nights a week.

In general, teachers thought that the mental computation, development, review, and homework phases of the program were best. Many noted a higher level of student interest in mathematics after implementation of the project. Several teachers appreciated the increased thought they gave to math instruction, particularly the developmental stage.

Negative responses to the program (in response to requests for information about weakest or most confusing parts) were infrequent, quite variable, and general in nature. Three teachers had difficulty getting pupils to hand in homework; three others complained that students initially had trouble with mental computation exercises which were unfamiliar to them. A few teachers thought the program was not flexible enough to allow for a wide range of student abilities within their classes.

In total, the affective reaction of teachers to the program was extremely positive. Their response indicated a general willingness to continue using the program and suggests that the program does not present an increased level of work that is apt to be unacceptable to the average classroom teacher.

Replication

Other investigators have reached similar conclusions about the generally positive effects of the program on students and teachers. For example, Keziah (1980), and Andros and Freeman (1981) have found in independent studies that teachers' affective reactions to the program are quite favorable. In these studies, teachers indicated anonymously that they found the program valuable and interesting. Teachers were also willing to continue using the program. Although we have attempted to obtain confidential follow-up information from teachers, it is always possible that teachers may tend to provide more positive evaluative responses (social desirability effects) when responding to program developers than they would to other professionals. The data presented by Andros and Freeman

(1981) and by Keziah (1980) strongly suggest that this is not the case with our program.

There are also data which indicate that the general program has increased the mathematics achievement of students in other settings as well. For example, in the Linn Mar Community School District (Marion, Iowa) we presented an initial description of the program and observed classrooms of several participating teachers months later, but did not monitor the day-to-day program implementation. Data provided by Dr. Harold Hulleman suggest that the program had very positive effects upon students' achievement and teachers' attitudes.

The participants in the Linn Mar program evaluation were 600 students and 24 teachers in grades four, five, and six. Students were pre- and posttested in September and May using the Stanford Diagnostic Mathematics Test. Students at all grade levels generally benefited from participation in the project. In comparison to expected pre/post gains, fourth-grade students' scores were five months higher and fifth- and sixth-grade students' scores were respectively seven and six months higher than expected. Teachers' affective responses to the program were also very favorable. For example, 18 of 20 teachers rated the program as "very good" (11) or "good" (7).

Experimentation elsewhere has also produced positive results (e.g., Milwaukee School Improvement Project, Maureen Larkin, Personal Communication) and the program is presently being tested in some schools in five different counties in Maryland (Cecil, Garrett, Harford, Montgomery, and St. Marys). The program is being used in other states as well. Evaluation of program effects in these projects is still in progress; however, informal feedback has been positive.

Some school districts have also modified the program to fit their instructional context. For example, Dr. Joe Dobson and Ms. Jane Timmerman have altered the program for use with remedial junior high mathematics students in the Winston-Salem/Forsyth County School District. The mathematics teachers spend at least three days per week in activities involving whole-class instruction using the Missouri model. The remaining two days per week are used for more individualized work in an attempt to meet the wide range of student abilities in secondary schools.

Trained observers found that the mean time-on-task percentage across several program classes and teachers was 70.5 percent. These data are particularly impressive when one considers the fact that these students were low-ability students who traditionally have difficulty staying on task. The 70.5 percent mean in program classrooms was also substantially higher than the mean for comparison classrooms composed of similar students where the mean percent of time on task was 53.5 percent. Achievement gains were also substantially higher in program classrooms. For example, in seventh-grade classrooms the mean gain for program students was 5.10

while control students' gains were 2.69 (normal curve equivalents). In eighth-grade classrooms the mean gains were 7.07 for program students and 3.84 for control students.

Other school districts are adapting the program for use in other school subjects (science). With the results from these and other projects we will have better information about the effects of the program in other settings and how modifications in the program facilitate or reduce the impact of the program on student achievement.

Program Adaptation: Two SES Populations

We also have attempted to determine if and how the program should be adjusted in different settings. Despite our initial, very positive results, we did not believe then, nor do we believe now, that there is one optimal way to present mathematics instruction that is equally effective for all teachers or students. Rather, we believe that the general principles of the program are generalizable to various types of students and different types of mathematics content. However, adaptations would have to be made for certain combinations of student and teacher types. In Chapters 7 and 8 we present discussions of the differential effects of the program in terms of derived teacher and student types. Here we want to discuss our study of teacher effectiveness across two different student populations.

In Chapter 2 we collected process data from a relatively affluent school district which served a middle to upper-middle-SES[1] population. In contrast, the data described earlier in this chapter were drawn from a population serving a lower-SES sample, where teachers' initial mean classroom achievement levels were below grade level.

There are a number of similarities between the two sample populations. For example, in both cases roughly 40 fourth-grade teachers were observed teaching mathematics at least six times between October and December. Furthermore, pre- and postachievement data were collected in each study, making it possible to compute a mean residual score for each teacher. These scores provide an operational definition of teaching effectiveness which can be related to instructional process measures.

It is important that a large set of common process measures was coded in both studies, thus making it possible to study the impact of identical teaching behaviors in two different SES contexts. Furthermore, the two coders who collected all of the data in the high-SES study were

1. It is important to recognize that SES is a proxy variable that stands for a complex set of factors. Obviously, there are both high and low achievers and students with favorable and unfavorable attitudes toward school in both samples. The samples should be viewed as different but overlapping populations. Proportionately, in one sample, the students came from moderate and moderately high-income families and achieved at moderately high levels. In contrast, in the other sample, relatively more students came from low-income families and initially were achieving at comparatively low levels.

members of a four-person coder team who collected data in the lower-SES study.

A more important consideration is that about half of the teachers in the low-income schools were part of an experimental study and were asked to teach in specified ways. Thus a treatment influence may have altered naturally occurring process/product relationships. However, the data are still relevant to the more generic question of *what processes appear to relate to achievement?*

Only a few of the process measures reported here were part of the explicit treatment study and called to teachers' attention. (Much of the treatment dealt with lesson stages and sequences of the lesson stages rather than individual process measures.) While the differences dilute the SES comparison somewhat, the strength of the comparison still seems evident.

The major process variables used for comparisons in this research were collected with the Brophy-Good Dyadic System (Brophy and Good, 1970). In both studies, observers reported over 80 percent agreement on all coding categories in the system. Because of differences in class period lengths (some teachers taught 37 minutes; others 45) all of the process measures were time-adjusted to represent estimates of occurrences per hour.

Residual scores were computed for each student on the achievement test by using the student's pretest score as a covariate. Data for teachers were then compiled by computing the mean residual score of their students.

Results

The most striking finding from the data presented in Table 4.5 was that few of the behavioral comparisons showed important differences across the two samples. However, some minor and major differences do appear. There are very minor differences in the types of questions that appeared to be most useful in the two settings. The correlations are quite low, but comparatively academically focused questions (product rather than self-reference or opinion questions) appeared to be important in the low setting. Simple product questions seemed to be more useful than process questions in both settings.

Another subtle difference appears in terms of teachers' reactions to students when they do not answer questions or answer incorrectly or incompletely. In the higher-SES setting, it was useful for the teacher to stay with a student and work for further response from the student who gave a partially correct answer. In contrast, it seemed better in the low-SES classroom to maintain momentum and not to continue working with the individual student in a public setting, even when a student gave a partially correct response.

TABLE 4.5 Correlation of Behavioral Measures with Teachers' Residual in Both the Low and High SES Classrooms

Variable	Correlation in low SES	p Value	Correlation in High SES	p Value
Classroom Climate[1]	0.42	0.01	0.28	0.11
Managerial[1]	0.10	0.57	0.00	0.97
Total Class Time	0.10	0.56	0.18	0.32
Transition Time	−0.27	0.12	0.11	0.55
Time Going Over Homework	0.26	0.14	0.02	0.91
Review Time	0.09[3]	0.60	0.29	0.10
Development	−0.14[3]	0.41	−0.13	0.50
% of Students Probably Involved	0.45	0.01	−0.07[1]	0.68
Student Asks Question	0.13	0.46	0.09	0.60
Discipline Type Question	−0.04[2]	0.79	−0.30[2]	0.08
Direct Question	0.16	0.34	−0.08	0.65
Open Question	0.15	0.39	0.17	0.64
Student Calls Out Answer	0.01	0.95	0.32	0.06
Process Question	−0.16[3]	0.34	−0.15	0.60
Product Question	0.15[3]	0.39	−0.02	0.89
Choice Question	0.14	0.42	0.15	0.59
Self-Reference Question	−0.07[2]	0.66	0.20[2]	0.25
Opinion Question	−0.25[2]	0.14	0.05[2]	0.76
Correct Response	0.11	0.51	−0.03	0.85
Partially Right Response	−0.09	0.60	−0.13	0.51
Wrong Response	0.16	0.36	−0.20	0.26
"Don't Know" Response	0.02[2]	0.88	−0.22[2]	0.20
No Response	−0.12	0.47	−0.19	0.28
Praise	0.35	0.04	−0.18	0.67
Affirm	0.04	0.80	0.15	0.57
Summarize	−0.39[2]	0.02	−0.13[2]	0.50
No Feedback	0.26[2]	0.13	0.03	0.84
Negate Wrong	0.15	0.39	0.04	0.80
Criticism	−0.09[2]	0.58	0.02[2]	0.91
Process Feedback	−0.11	0.53	0.27	0.12
Gives Answer	−0.09	0.59	−0.02	0.90
Ask Another Student	0.30	0.08	−0.07	0.70
Another Student Calls Out Answer	−0.20[2]	0.26	−0.15[2]	0.59
Repeats Question	−0.25	0.15	−0.07	0.68
Gives Clue	−0.06	0.73	0.03	0.87
Asks New Question	−0.28	0.10	0.12	0.51
Expands Student's	−0.32[2]	0.06	0.02[2]	0.92

Variable	Correlation in Low SES	p Value	Correlation in High SES	p Value
Response				
Student Initiated Work-Related Contact—Teacher Praise	0.12^2	0.47	0.28	0.11
Student Initiated Work-Related Contact—Teacher Gives Process-Type Feedback	−0.25	0.15	0.14	0.56
Student Initiated Work-Related Contact—Teacher Gives Feedback	0.00	0.99	0.37	0.03
Student Initiated Work-Related Contact—Teacher Criticizes	-0.37^2	0.02	-0.11^2	0.55
Student Initiated Work-Related Contact—Teacher Type Feedback Unknown	0.04^2	0.78	-0.21^2	0.24
Right Response Followed by Teacher Praise	0.35	0.04	0.09	0.62
No Response or "Don't Know" Response Followed by Sustaining Feedback	-0.22^2	0.20	−0.21	0.24
No Response or "Don't Know" Response Followed by Terminal Feedback	0.01	0.93	−0.22	0.23
Wrong Response Followed by Terminal Feedback	0.22	0.21	−0.19	0.29
Wrong Response Followed by Sustaining Feedback	−0.16	0.36	−0.10	0.56
Part Right Response Followed by Terminal Feedback	-0.02^2	0.88	0.11	0.58
Part Right Response Followed by Sustaining Feedback	-0.18^2	0.29	0.30	0.12
Total Response Opportunities	0.16	0.36	0.14	0.55
Total Teacher Initiated	0.28	0.10	−0.33	0.06

Variable	Correlation in Low SES	p Value	Correlation in High SES	p Value
Work-Related Contacts				
Total Teacher Initiated Behavior-Related Contacts	0.00	0.97	−0.21	0.23
Total Teacher Initiated Contacts	0.22	0.20	−0.33	0.06
Total Student Initiated Work-Related Contacts	−0.06	0.70	0.37	0.03
Total Student Initiated Procedure-Related Contacts	0.10^2	0.55	−0.03	0.87
Total Student Initiated Contacts	−0.05	0.75	0.35	0.04
Total Dyadic Contacts (Student Initiated, Teacher Initiated, and Response Opportunities)	0.04	0.79	0.17	0.66
Direct Question Direct, Plus Open Question	0.00	0.98	0.05	0.76
Direct Question Response Opportunities	0.12	0.48	0.02	0.91
Open Questions Response Opportunities	0.09	0.61	−0.09	0.62
Call Outs Response Opportunities	−0.07	0.69	0.00	0.98
Student Initiated Work-Related Contacts Total Student Initiated Contacts	−0.01	0.93	0.01	0.96
Teacher Initiated Work-Related Contacts	0.32	0.06	−0.26	0.13
Total Teacher Initiated Contacts Total Student Initiated Contacts	0.11	0.53	−0.34	0.05
Process Questions Total Questions	$−0.31^3$	0.07	−0.19	0.28
Choice Questions	−0.06	0.70	−0.25	0.15

Variable[4]	Correlation in Low SES	p Value	Correlation in High SES	p Value
Total Questions Opinion Questions	−0.16	0.35	−0.03	0.84
Total Questions Product Questions	0.25[3]	0.14	−0.10	0.60
Total Questions Correct Responses	0.00	0.96	0.25	0.15.
Total Responses Wrong Responses	0.21	0.22	0.19	0.27
Wrong Responses Plus No Response "Don't Know"	0.04	0.81	−0.16	0.61
"Don't Know" Plus No Response % of Responses Teacher Gave No Feedback	0.08	0.64	−0.07	0.71
Student Initiated Procedure-Related Contact—Teacher Praise	0.16[2]	0.35	0.18[2]	0.30
Student Initiated Procedure-Related Contact—Teacher Gives Feedback	0.10[2]	0.56	−0.05	0.76
Student Initiated Procedure-Related Contact—Teacher Criticizes	−0.02[2]	0.87	−0.10[2]	0.57
Teacher Initiated Work-Related Contact—Teacher Gives Praise	0.18[2]	0.29	−0.14[2]	0.56
Teacher Initiated Work-Related Contact—Teacher Gives Process Feedback	0.24	0.16	−0.29[2]	0.10
Teacher Initiated Work-Related Contact—Teacher Gives Feedback	0.22	0.20	−0.25	0.15
Teacher Initiated Work-Related Contact—Teacher Criticizes	−0.06[2]	0.70	−0.19[2]	0.30
Teacher Initiated Work-Related Contact—Teacher	0.16[2]	0.36	−0.28[2]	0.11

Variable	Correlation in Low SES	p Value	Correlation in High SES	p Value
Type Feedback Unknown				
Teacher Initiated Behavior-Related Contact—Teacher Gives Procedure Feedback	-0.10^2	0.57	0.02	0.88
Teacher Initiated Behavior-Related Contact—Teacher Praises	0.05^2	0.78	0.05^2	0.79
Teacher Initiated Behavior-Related Contact—Teacher Warns Student	0.02	0.89	−0.30	0.08
Teacher Initiated Behavior-Related Contact—Teacher Criticizes Student	-0.18^2	0.30	-0.05^2	0.76
Wrong Response Followed by Teacher Criticism	-0.11^2	0.51	−0.15	0.61
Process Feedback Response Opportunities	−0.01	0.95	0.24	0.17
Process Feedback Product Feedback	−0.01	0.93	0.25	0.15
Expands Feedback Total Feedback	−0.24	0.16	−0.15	0.57
Process Feedback in Student Initiated Work Related Contacts Total Student Initiated Work	−0.17	0.34	−0.19	0.30
Process Feedback in Teacher Initiated Work Related Contacts Total Teacher Initiated Work Related Contacts	0.14	0.43	−0.20	0.25
Total Process Feedback	−0.10	0.54	0.16	0.61

1. High-inference variable.
2. Correlations based on variables with a low frequency of occurrence.
3. Correlations that might be contaminated by the treatment.

4. Most of the variable descriptors are self-explanatory; however, some may need additional clarification as provided below. For an extended discussion of these definitions and coding examples, see Brophy and Good (1970).

DIRECT QUESTION: Teacher calls on a child who is not seeking a response opportunity.

OPEN QUESTION: The teacher creates the response opportunity by asking a public question, and also indicates who is to respond by calling on an individual child, but chooses one of the children who has indicated a desire to respond by raising his/her hand.

PROCESS QUESTION: Requires students to explain something in a way that requires them to integrate facts or to show knowledge of their interrelationships. It most frequently is a "why?" or "how?" question.

PRODUCT QUESTION: Product questions seek to elicit a single correct answer which can be expressed in a single word or short phrase. Product questions usually begin with "who?", "what?", "when?", "where?", "how much?" or "how many."

CHOICE QUESTIONS: The child does not have to produce a substantive response but may instead simply choose one of two or more implied or expressed alternatives.

SELF-REFERENCE QUESTION: Asks the child to make some non-academic contribution to classroom discussion ("show and tell," questions about personal experiences, preferences, or feelings, requests for opinions or predictions, etc.).

OPINION QUESTIONS: Much like self-reference (except no one correct answer) except that they seek a student opinion on an academic topic (e.g., Is it worth putting a man on the moon?).

NEGATION OF INCORRECT ANSWERS: Simple provision of impersonal feedback regarding the incorrectness of the response, and not going further than this by communicating a personal reaction to the child. This can be communicated both verbally and nonverbally.

Praise seems to have an important but differential impact in the two settings.[2] In the high-SES classroom, praise was negatively related to student achievement, but was positively related in the low-SES setting. The climate variable shows that a relaxed, more pleasant learning atmosphere is facilitating in both settings, but a relaxed climate appeared to be more important in the low-SES setting.

The management of seatwork is another area where differences emerged. In the high-SES setting, it appeared desirable for teachers to allow students to seek them out (student-created, work-related contact); however, teacher-initiated contact seemed to be more strongly associated with student achievement gains in the low-SES setting.

2. The dysfunctional effects of praise in the higher-SES sample are shown more fully in an analysis of the top and bottom nine teachers (Good and Grouws, 1977). This analysis shows that teachers who got the best achievement used significantly less praise than teachers who got the lowest results.

Also, it should be noted that student involvement codes were strongly related to achievement in the low-SES sample, but virtually no relationship existed in the high-SES sample.[3]

We have found four context differences in the present study and two of these seemed important. Low-SES classrooms demand that teachers supervise and monitor students' seatwork actively and control private interactions with students (decide which student to contact rather than allowing students to seek them out). A second finding is that positive affect and a relaxed climate are more important in low- than higher-SES settings.

Both of these context effects were also reported by Brophy and Evertson (1976); hence, this replication makes these data much more generalizable. Also, both of the findings are reasonably consistent with findings in developmental and social psychology. Low-achieving students, compared to high achievers, have shorter attention spans and are more distractible. Hence, teacher contact (or the physical nearness of the teacher) and feedback help students to maintain task involvement. The low achievers' greater need for positive affect, and the damaging impact of negative teacher affect, can be explained in terms of low-achievement students' previous history of failure in the classroom. Such students are likely to interpret neutral or ambiguous feedback as an indication of failure. Miller (1975) reports data that seem to imply that minority students have a greater need for approval. He reported that minority students are likely to derogate their own ability in failure situations and to show less tolerance for conflict (Miller, 1970).

Two weak context findings also appeared in the data. The first of these, that low achievers seemed to benefit from product questions, is also a finding that Brophy and Evertson report. However, the data presented here suggest that the questions should also be academically focused. That is, questions about subject matter appeared to be more useful than self-reference questions (e.g., What do you like to do after school?). Frequent use of nonacademic questions may be an expression of low teacher expectations and may divert student attention from the major, substantive discussion. However, it is difficult to attach much importance to this finding since both self-reference and opinion questions occurred infrequently.

The other context finding, that more effective teachers tended not to stay with low-SES students in failure situations, is in direct conflict with the Brophy-Evertson data. They reported that more effective teachers in low-SES classrooms tended to stay with (repeating or rephrasing the ques-

---

3. It should be noted that involvement was coded as a high-inference variable (measured on a rating system) in the high-SES sample, but coded as a low-inference variable (actually counted) in the low-SES sample. This methodological difference might be a major part of the differential results.

tion) students in failure situations. These conflicting results will only be resolved with additional research. However, some differences in the two studies may be important.

One plausible explanation for the differences between the two studies is the exclusive focus on mathematics in the present study. Given that the correctness of the answer is often more verifiable in math than in other subjects, it may be that students have greater self-evaluation capacities in mathematics. That is, their failure to respond in mathematics is because they don't know the answer (rather than because of their anxiety about speaking publicly); hence, it may make more sense to deal with students' misunderstandings privately than publicly. Also, students in the Brophy-Evertson study were younger (second and third graders) and may have had less capacity for determining whether or not they knew an answer.

The fact that ratings of student involvement correlate positively in the low-SES setting but not in the high-SES setting is probably due to the fact that perceived student involvement is more important in a low-SES setting. Immature learners probably cannot attend to two or three things at the same time. They may not have the capacity to look out the window and watch other children on the playground and still listen to the teacher as more mature learners can. Part of the difference can probably be explained in terms of methodology. In the low-SES sample, student involvement was actually counted. However, in the high-SES sample it was estimated with a high-inference code. Still, we suspect that student involvement may be a better proxy for student learning in lower- than in high-SES settings.

One of the difficulties of research using mean classroom achievement as a criterion is that prescriptive statements such as "it makes sense" are restricted to effects on the class as a whole. What makes sense for a given student or subgroup of students may be detrimental to the class as a whole and vice versa (Good and Power, 1976). Effectiveness research has failed to deal with the subgroup–whole-class issue and this seems the next important step in the research paradigm.

Teaching is complex and invariably teachers have to adjust their teaching to the particular group of students in their class. No pattern of teaching is going to apply uniformly in any setting. For example, on a probability basis it appears that low-SES students need more praise than do high-SES students. However, it is equally true that some high-SES students will benefit from teacher praise and that some low-SES students do not need high levels of teacher praise.

In general, the data presented here suggest that SES differences are real but perhaps not as great as suggested by Brophy and Evertson (1976) and Medley (1979). However, the reader should realize that our data may represent a "minimum case" of SES differences, since student age (fourth graders) and subject matter (mathematics) were controlled.

Other age levels and/or subject areas, such as reading or social studies (which may be more affected by social influences than mathematics), might show more extreme differences. Additional research at various grade levels and in different subject areas is needed to answer these questions.

The comparison between the two school populations provided some clues about possible adaptations in instructional style that may be necessary for the program to be effective in different contexts. Still, both data sets suggest that relatively active and systematic teaching is associated with higher student achievement. We will discuss other research efforts to determine information about how to adapt the program in Chapters 7 and 8.

DISCUSSION

Given the short period of the treatment program and the relative ease of implementation, the results of this study are important. It is part of a recent trend (e.g., Anderson, Evertson, and Brophy, 1979; Stallings, 1980) in research on teaching that is beginning to show that not only do well-designed process/outcome studies yield coherent and replicable findings, but treatment studies based on them are capable of yielding improvements in student learning that are practically as well as statistically significant. Such data are an important contradiction to the frequently expressed attitudes that teaching is too complex to be approached scientifically and/or that brief, inexpensive treatments cannot hope to bring about significant results.

Also, it is important to note that these gains are made in urban, low-income schools. That achievement increments can occur in such schools is aptly demonstrated by this project, and this experimental finding appears to be important, given the low expectations that many educators hold toward inner-city schools. It is also the case that increases in achievement came by influencing *how* teachers used time rather than by increasing the time allocated for mathematics.

It is interesting to note that the study had positive effects on both control and experimental teachers. That control teachers and their students showed marked improvement was probably due to the strong Hawthorne effect that was purposefully built into the project. Increased motivation probably led control teachers to think more about their mathematical instruction, and such proactive behavior (e.g., more planning) may have brought about increased achievement. The presence of a strong Hawthorne control makes it possible to argue with more confidence that the resultant differences between control and treatment classes were due to the instructional program and not to motivational variables. The positive findings of researchers who have examined the program in other settings also provide support for the claim that the program is generally effective.

We are *not* suggesting that the instructional program used in the study is the best approach to take for facilitating the mathematics achievement of students. However, we are arguing that it appears to have considerable value for teachers who utilize and/or prefer a whole-class organizational pattern for teaching mathematics in the middle elementary grades. Although students at this age appear to benefit from the program, all their mathematics instruction should not necessarily be of this mode. A comparison of teaching in relatively high- and low-SES settings indicated some areas where teaching adjustments appear to be necessary. Still in both settings it appeared that active and systematic teaching was associated with greater student achievement.

As noted previously, some of the individual instructional behaviors correlated moderately highly and positively with student achievement; however, it must be emphasized that these behaviors occurred along with other variables. For example, students did homework only after they had been shown how to do the assignment in teacher-supervised seatwork activity. Hence, it is difficult and perhaps misleading to overemphasize the meaning of any individual behavior. At this point the most reasonable interpretation is that the total instructional treatment program, when implemented, had a positive impact upon mean student achievement. The importance of a particular variable can only be evaluated in subsequent studies that delete certain other variables in the instructional program. Such work may allow for an understanding of program components that appear to be most strongly related to achievement gains. Such analysis can help determine which behaviors to delete or modify in subsequent studies.

Development is one variable that would seem to need clarification in future research. Our research demonstrated the need to improve the observation scale for development. This could involve trying to pinpoint behaviors that characterize development and improving the quantitative measures of this component. Another appropriate direction to pursue is the creation of reliable assessments of development along qualitative dimensions. A more detailed statement about development will be presented in Chapter 9.

Continued efforts to improve and refine the entire treatment program are necessary if more insight into the teaching of mathematics is to be achieved. Still, the magnitude of the treatment effect is important and offers convincing proof that it is possible to intervene successfully in school programs. These data are consistent with other recent treatment interventions in elementary schools (Good, 1979; Good, 1980; Good, in press).

Finally, it should be noted that although students and teachers who were exposed to the Missouri Mathematics Program in this experiment generally reflected higher achievement than did control classrooms, some combinations of teachers and students appear to especially benefit from the program. These findings and the work of others that have examined the differential effects of instructional programs on different types of teachers and students will be presented in Chapters 7 and 8.

5

Elementary School Experiment II

Based upon our first experiment (reported in the previous chapter) we were interested in seeing if we could have a similar positive effect upon students' mathematics abilities in sixth-grade classes. However, we also were concerned about enhancing students' abilities for doing verbal problems. This interest was motivated by the fact that the verbal problem-solving scores of the treatment students did not appear much different from scores of control students on the content test.

Dr. Robert Reys had designed a special content test for the first experimental study. The reliability of the instrument as a whole was good and showed that experimental students' achievement was superior to that of control students. The reliabilities for the three subtests of the instrument (knowledge, skill, and problem solving) showed that only the skill subtest had adequate reliability for separate analysis (and on this subtest the achievement of the treatment group surpassed that of the control group). However, in examining the means of the other two subtests, we found that treatment students appeared to do better than controls on the knowledge items but that there was little difference between the two groups on the verbal problem-solving test. Obviously, it was impossible to tell whether the comparability of the two groups was real or only a function of poor reliability (e.g., too few items).

We were disappointed in this aspect of the findings because we felt that if mathematics knowledge is to be applied to "everyday" matters, students need skills in this area (e.g., to compare whether the 12 oz. or 16 oz. package is the better buy). Unfortunately, the extant literature on instructional behavior and students' performance on verbal problem solving did not lead to any consistent orientation or procedure. We felt that it was important to understand and to possibly improve students' ability for solving relatively simple verbal problems. We thus decided to make a systematic effort to develop testable instructional strategies in this area.

Treatment Program for Study II

The first task was to develop a training manual suggesting instructional strategies that teachers might use to influence students' verbal problem-solving skills. The five techniques that teachers were requested to use were problems without numbers, writing verbal problems, estimating the answer, reading verbal problems, and writing open-sentence problems. Discussion of these strategies and related research can be found elsewhere (e.g., Suydam and Weaver, 1970). Space limitations prevent an extended discussion of the rationale and procedures presented in the training manual, but one brief example follows to provide some understanding of our procedural directions. The entire training manual appears in Appendix B.

Problems Without Numbers

The use of problems without numbers is one instructional technique for improving verbal problem-solving performance (Riedesel, 1964). It provides students an opportunity to gain insight into the problem-solving process by avoiding the use of numbers and thus the need to perform any computation whatever. To illustrate the method, consider the following typical problem:

> Two classes sold 100 football game tickets.
> One class sold 27 tickets.
> How many did the other class sell? (Holt School Mathematics, Grade 6, p. 32.)

This problem can easily be rephrased so that it is a problem without numbers:

> Our class and Mrs. Smith's class sold tickets. We know how many tickets were sold altogether and how many tickets our class sold. How many tickets did Mrs. Smith's class sell?

The teacher presents only the problem without numbers and asks the class *how* to solve it. An appropriate answer might be something like this: "I'd subtract how many tickets we sold from the total number of tickets to find how many tickets Mrs. Smith's class sold." Time permitting, the teacher should follow up with another problem without numbers, or occasionally consider the same problem, but with the numbers included.

Rationale

This technique may be effective because it causes students to focus extensively on the *method* needed to solve a problem without any numerical or computational distractions. Many teachers realize that too frequently students begin doing computation before they have really thought through a problem. In fact, some students have been known to begin computing before they have read the entire problem! Avoiding the use of numbers tends to resolve these kinds of problems. Since the strategy does not require computation, students can be exposed to a substantial number and variety of verbal problems in a short period of time.

Other Experiment II Decisions

After having made the decision to shift our focus to problem solving, it was also necessary to make three related decisions: (1) whether to test the instructional materials associated with verbal problem solving with or without the program that had been designed for the first field experiment; (2) at what grade level(s) to test the program; and (3) whether to observe or not.

It seemed more reasonable (at the time) to see if the previous gains could be maintained in knowledge and skill areas while also improving students' performance on verbal problem-solving skills. Because the program's effectiveness had been demonstrated within the context of that school program, it seemed more reasonable to test an expanded, comprehensive program rather than to test only a piece of the program.

The grade-level decision was a relatively straightforward one. We could have tested the program at the fifth-grade level and thereby have gained the advantage of looking at students over consecutive years. However, the movement from school to school within the student population was relatively high. Student movement would mean that some teachers would have some fifth-grade students who had been in the program as well as some students who had not. To avoid this confusion, we decided to test the modified program at the sixth-grade level. We could thus test the program on an "uncontaminated" population of classrooms, and we would also have an older population upon which to test various questions about the program (e.g., does it have too much structure for older students?).

The final decision we had to make concerned the role of observation in the field experiment. Limited funds, the fact that new observers had to be trained, and our interest in building a new treatment program (as well as exploring the previously collected typology data), collectively influenced the decision of whether to observe or not. Limited observation was a possibility.

Our decision *not* to observe was based on a realization that if observation remained a part of the treatment, the successful application of the program would be limited to situations where repeated classroom observation was included. Classroom observation is an expensive item and we were curious to see if the program would work without it.

Our interest in testing the program without observation was stimulated by our awareness that observed and unobserved treatment teachers were both successful in obtaining student achievement gain in the field experiment conducted by Anderson, Evertson, and Brophy (1979). Their results suggest that the treatment had an effect upon student achievement which was not moderated by the presence of observers.

Our own observational data had suggested that most of the experimental teachers implemented the program reasonably well and that some parts of the program were better implemented than others. In general, those behaviors that were implemented most consistently involved specific requests and required no extra work on the part of the teachers. It seemed to us (at the time) that the new teaching requests being made were relatively specific although we realized that implementation might involve some extra work. Ultimately, we decided not to observe teachers and to see if students' achievement in several areas of mathematics could be improved without elaborate training (e.g., videotapes) or classroom observation.

Unfortunately, despite our efforts to secure an "uncontaminated" population by avoiding fifth-grade classrooms, a degree of contamination was present in the design. In part, we were "victimized" by success. The school district was sufficiently impressed with the results of the first study that they wanted all fourth-grade teachers to be exposed to the model. Due to this dissemination (which we helped with), as well as our own debriefing of control teachers, program descriptions of the first experimental treatment were present in most schools and thus potentially available to sixth-grade teachers. In retrospect, it might have been more profitable to have only studied the problem-solving materials (which were available only to sixth-grade teachers) in the second experiment. Without observation, it was impossible to determine whether sixth-grade control teachers were aware of and/or using parts of the treatment program.

Method for Experiment II

The design for this experiment was similar to that used in the first field study. The expanded program (general program plus verbal problem-solving manual) was evaluated in 36 sixth-grade classrooms. Schools were matched on the basis of SES and then randomly assigned to treatment and control conditions. Three organizational patterns were present in the sample: the semidepartmentalized structure (which was utilized exclusively in the first field study and which has been explained previously); math-as-a-special-subject (these sixth-grade teachers taught math to several different sixth-grade classes); and open classrooms (where team teaching and individualized instruction were prevalent).

The semidepartmentalized structure and math-as-a-special-subject organizational patterns seemed to be consistent with the basic data base from which the project had been developed. The open-classroom structure was not. However, the school district expressed interest in including some of these classrooms in the design in order to have teachers exposed to the rationale for the directed teaching aspects of the program. We included these teachers in the design but emphasized that the treatment would be conceptual rather than operational (if certain aspects of the program stimulated interest, the adaptation would be left to the teachers). Our plan was to analyze program results twice, with and without open teachers included. The final sample for the treatment group was: math as a special subject (5); semidepartmental structure (9); open classroom (3). For the control group the final sample was: math as a special subject (5); semidepartmental (10); and open structure (4).

Teacher Training

On October 1, 1978, we met with all teachers participating in the project. The project's focus on the improvement of student achievement in mathematics was explained. Again, key officials from the school district were present (to create the Hawthorne conditions described in the first experiment) and the interest of the school district was communicated. Teacher descriptive data and forms were distributed and completed. At this point the treatment and control teachers were divided into separate groups. Control teachers were informed that they would receive delayed feedback and that their role in the project was to teach as they had been. Treatment teachers were presented with the philosophy and details of the instructional program (1½ hours). Experimental teachers were also given the 45-page general manual and the 17-page problem-solving manual. Teachers were asked to implement the basic program and also to spend 10 minutes a day on verbal problem-solving strategies. Two weeks after implementation of the treatment had begun, another meeting was held with treatment teachers to answer questions and to resolve any problems associated with the treatment (1½ hours). Total contact with the treatment teachers was three hours.

Testing

Students were administered the mathematics subtest of a standardized achievement test (Science Research Associates; SRA, Short Form E, blue level) at the beginning of the project (early October) and as a posttest in mid-February. Additional postmeasures included a 20-item re-search-constructed verbal problem-solving test, basically the same student typology instrument that had been used in the first field experiment (see Appendix A), and a short attitude scale (the last 10 questions of the student typology instrument).

RESULTS OF EXPERIMENTAL STUDY II

The raw means and standard deviations for the SRA pre- and posttests and the problem-solving posttest are presented in Table 5.1, by treatment condition and by organizational structure. As can be seen in all three types of classrooms, student performance increased from pre- to post- on the 40-item SRA test. Furthermore, the treatment group surpassed the performance of the equivalent control group in all cases. In terms of per-formance on the problem-solving test, two of the three treatment groups had higher mean performances than did the equivalent control group. It should be noted that the exception, the open-treatment classes, had the lowest pretest scores on the SRA.

As can also be seen in Table 5.1, the mean pretest SRA scores for control teachers were generally lower than was the case in the equivalent treatment group. The only exception occurred in the math-as-a-special-subject classes, where the pre-SRA mean scores of the treatment classes slightly exceeded those of control classrooms. To reiterate, in terms of raw gains, it was found that the treatment group's performance was generally superior to that of the equivalent control group.

What, then, were the effects of the formal analyses using adjusted mean scores? To pursue this question we used teacher classroom means

TABLE 5.1 Means and Standard Deviations on Pre- and Post-SRA and
Post-Problem–Solving
Test by Instructional Group and by
Classroom-Organization, Field Experiment II

| | Pre-SRA | | Post-SRA | | Pre-Post Change on SRA | Post-problem | |
	$\bar{x}$	SD	$\bar{x}$	SD		$\bar{x}$	SD
Control	26.80	4.1	29.65	3.7	2.85	14.71	1.6
Semi	27.35	4.1	30.56	4.0	3.21	14.86	1.8
Open	25.36	2.7	27.70	2.6	2.34	14.55	.85
Special	27.26	5.9	29.78	4.3	2.52	14.56	2.1
Treatment	25.03	5.0	28.96	4.8	3.93	14.90	2.0
Semi	25.22	4.2	28.71	4.8	3.49	15.17	1.4
Open	20.41	1.3	26.01	4.9	5.60	13.13	3.6
Special	27.44	6.3	31.18	4.7	3.74	15.46	1.8

because we felt that this was the most appropriate form of analysis. In comparing student performance on the post-SRA test, using the pre-SRA test as the covariate (with all forms of classroom organization included in the analysis), it was found that the performance of the treatment group was not significantly higher than that of the control group ($p = .26$). When these data were reanalyzed using the student as the unit of analysis, the p-value was found to be lower ($p = .13$). However, we stress that the group mean analysis is more appropriate. We include this one example with student unit data to illustrate that this type of analysis yields a more favorable interpretation, but one we feel is erroneous.

A similar analysis was performed on the problem-solving test (using the pre-SRA as a covariate), to compare the significance of adjusted means across all treatment and control classrooms (using classrooms as the unit of analysis). This analysis reflected that the performance of the treatment group exceeded that of the control group in a way that approached significance ($p = .10$).

Earlier it was mentioned that we had some reservations about including open-classroom teachers in the study because the program had not been designed for such settings. When the analyses were repeated without open class teachers (using the class as the unit of analysis and the pre-SRA as a covariate), it was found that the performance of treatment classrooms on the SRA posttest did not significantly exceed that of the control group. However, the comparison on the problem-solving test revealed that the treatment group's performance was significantly superior to that of the control group (.015). The source tables for these analyses and the adjusted means are presented in Tables 5.2 through 5.4.

TABLE 5.2   Analysis of Variance Results on Adjusted Mean* Post-SRA Test Scores (Using Pre-SRA Scores as a Covariate) between All Treatment and Control Classrooms

Source	DF	MS	F	Probability
Treatment condition	1	5.58	1.31	.26
Error	33	4.27		

*Note the adjusted mean for the control group was $-.38$ and for the treatment group .42.

TABLE 5.3   Analysis of Variance Results on Adjusted Mean* Problem-solving Scores (Using Pre-SRA Scores as a Covariate) between All Treatment and Control Classrooms

Source	DF	MS	F	Probability
Treatment condition	1	4.33	2.86	.10
Error	33	1.51		

*Note the adjusted mean for the control group was $-.33$ and for the treatment group .37.

TABLE 5.4   Analysis of Variance Results on Adjusted Mean* Problem-solving Test Scores (Using Pre-SRA Scores as a Covariate) between Treatment and Control Classrooms with Open Classes Dropped, Field Experiment II

Source	DF	MS	F	Probability
Treatment condition	1	5.45	6.77	.015
Error	25	.81		

*Note the adjusted mean for the control group was $-.45$ and for the treatment group .45.

Student Affect Data

Student affect, as measured by the ten-item affect test, was comparable before and after the treatment. The pre- mean for the control group was 17.83 and the pre- mean for the treatment group was 18.29. At the end of the experimental program, the mean for the control group was 18.45 and the treatment group's mean was 18.57. These data suggest that the affective reaction was similar for both groups and that the treatment had *no* meaningful impact on student attitudes.

Teacher Response

We assessed the reactions of the treatment teachers to the program in a confidential fashion, two months after the program had ended (teachers were given an unmarked response sheet to return by mail). In general, their responses indicated positive acceptance of the program and an intention to continue using it. Sixteen of the 17 treatment teachers responded to the survey.

The overall affective reaction of experimental teachers ($N = 16$) to the program was extremely positive. Thirteen teachers felt that all eight phases of the project were either "very good" or "good" (valuable to them as teachers). The regular assignment of homework and the review proved to be the most useful methodologies, while increased pace invoked the lowest affective response. Even so, 13 of the 16 teachers thought that the increased pace stage of the project was either "good" or "very good."

Questionnaire responses revealed that about two-thirds of the participants continued using all aspects of the program on or near the initial level recommended by the project directors. After the program ended, 10 teachers were still including verbal problem solving in their curricula, and 13 were implementing the prescribed development phase at least three times a week. Fifteen teachers continued to assign homework a minimum of three nights a week, and 13 were conducting weekly and monthly review sessions. In general, teachers thought that the development, verbal problem solving, review, and homework phases of the program were best. When asked about their negative responses to the program (weakest or most confusing parts), five teachers said they had difficulty using it with classes in which there was a wide range in student ability. Some of these teachers thought the program was particularly difficult for their low-ability pupils. Six teachers thought that there was not enough time allotted on a daily basis to complete all phases of the program.

Responses of Control Teachers

At the debriefing session we provided control teachers with a copy of the program manual. Two months later we assessed their reaction to these materials. We wanted to see how teachers who had been exposed to the program but who had not used it would react. Were the favorable comments of experimental teachers due to the fact that they had used the program and felt committed to recommend it? Also, we wanted to see how new various aspects of the program were to the control teachers. Their responses indicated that they were familiar with most parts of the material and in a couple of cases the free responses of the control teachers indicated that supervisors had advocated the use of the directed lesson to them.

Seventeen of the 19 control teachers responded to the questionnaire. Five of the 17 control teachers who responded to the questionnaire reported that they had carefully read both the general manual and the verbal problem-solving manual. Five others had read both manuals quickly, and six had at least skimmed them quickly and thought about the highlights. Responses revealed that there was considerable correspondence between the teaching methods control teachers were already using and those requested by the program. Eight teachers were already utilizing the prescribed development and seatwork aspects of the program, and were also teaching their classes as a whole. At least five more teachers reported general overlap between the program and what they had been doing, for each category except the verbal problem solving. This is of special interest because it was in the area of verbal problem-solving skills that treatment students were found to outperform control students. In general, control teachers said that the program was not new to them, although three teachers reported that the mental computation was somewhat novel. Two teachers had not been using the verbal problem-solving strategies previously.

Mental computation was listed as a strength of the program by three of nine teachers who responded to this item, and four teachers thought that the review sessions were especially useful. Only three teachers listed weaknesses of the program: it was hard for low achievers; there was not enough time to complete all parts of the program daily; and it was hard to get pupils to do homework on a daily basis. Five teachers planned to continue using the verbal problem-solving strategies outlined in the program, and four expected to continue the mental computation exercises. Two teachers said they would use the review, and two more planned to continue using the entire program.

Discussion of Experiment II

The reception of the program by the experimental teachers was excellent and this is an important result per se. Programs that increase student achievement, but which fail to gain teacher acceptance, probably have little durability. The fact that teachers indicated that they used and intended to continue using the program is gratifying, but the extent to which this verbal response matches actual teacher behavior is unknown. At a minimum, the program seems acceptable to teachers.

The achievement data indicate that both treatment and control groups made measurable progress. In terms of performance on the SRA posttest, the raw achievement gains of treatment classes surpassed those of control classes, but these results were not significant. This finding was not expected; in fact, based on the fourth-grade study, we expected a large effect. The lack of a significant effect in this study may be due to several factors. First, the fact that control teachers reported using many of the treatment behaviors may have diluted some of the treatment effect. Second, the treatment may not have been well implemented in all treatment classrooms. Perhaps because of the addition of the problem-solving strategies the teachers may have found the program overwhelming and hence, implemented only certain aspects of it. However, it should be noted that treatment teachers reported a high level of usage. Also, it could be that older students need less of the general treatment than do younger students.

The effects of the program on students' verbal problem-solving abilities were much more encouraging. Treatment classes were found to perform better than control classrooms on the verbal problem-solving test. These results were particularly pronounced when open classes were dropped from the analysis. Hence, the program provides strong support for the contention that instructional strategies can be utilized to improve student performance on verbal problems.

Follow-Up Experiment

Fortuitously, we had the chance to collect some follow-up data relevant to the verbal problem-solving component one year later.

Dr. John Engelhardt, an observer in our junior high study (see Chapter 6), completed his dissertation research examining the impact of the verbal problem-solving treatment upon students' performance in sixth-grade classrooms. As noted above, it had been shown that sixth-grade students' verbal problem-solving abilities were enhanced by exposing their teachers to the general Missouri Mathematics Program and to the verbal problem-solving treatment. The present study tested the verbal problem-solving treatment without the presence of the general treatment program.

As in all research projects, there is a need to limit the scope of the study so that certain questions can be addressed. Thus, certain questions cannot be examined. The issue in the present investigation was whether to vary program treatment (allow some teachers to use only the verbal problem-solving materials and assign other teachers both the general Missouri mathematics materials) or to vary observational format (observe some teachers and not others). Two major competing explanations for the results presented earlier in the chapter are (1) better results would be obtained by asking teachers to use but one of the programs (to do both immediately may be asking for too much), and (2) observation is necessary if certain aspects of the treatment program are to be used in the classroom. It would, of course, have been nice to have varied both conditions (as well as other factors!) but the resources and size of the sample available made it necessary to address one question. Ultimately, it was decided to pursue the effects of observation on program implementation, and observational format was varied while program format was held constant (experimental teachers were asked to use only the verbal problem-solving materials).

The following is a brief summary of the verbal problem-solving study and the interested reader can obtain detailed results elsewhere (Engelhardt, 1980). The study was undertaken to experimentally test the effectiveness of a program of systematic instruction in verbal problem solving on the achievement of sixth-grade students. The systematic instruction encompassed a daily time component of ten minutes (except on days when verbal problem solving was the main focus of the lesson) and five instructional strategies to be used in teaching problem solving—using problems without numbers, writing verbal problems, estimating answers, reading verbal problems, and writing open sentences. (The entire verbal problem-solving manual appears in Appendix B.)

The investigation was designed to answer the following questions:
1. Would the treatment increase problem-solving achievement?
2. Would the observation influence problem-solving achievement?
3. Would the treatment differentially affect the achievement of various student groups within the class?
4. Would the degree of treatment implementation correlate positively with residual achievement scores?

5. Would student attitude be affected?

Half of the teachers were observed in order to see how well treatment teachers implemented the teaching requests and to measure the extent to which control teachers dealt with problem solving. Observing half the treatment teachers also allowed for an examination of the impact of observation upon student achievement.

After initial instruction during an orientation workshop, the teachers in the treatment group were responsible for maintaining the problem-solving program in their classrooms. The teachers' main references were the verbal problem-solving manual and a procedure summary. Beyond these, teachers were to generate the resources necessary for the instructional program. At the conclusion of the experiment all teachers administered a problem-solving test and an attitude scale.

Since no pretests were administered, district data on file from spring 1979 testing were used as covariates in the analyses. Those students for whom complete data were on file were considered in the statistical analyses.

Results and Conclusions

With respect to the questions under investigation the following results were noted:

1. The treatment did not make a difference in problem-solving performance either on routine or nonroutine problems. The adjusted mean for the control group was higher than that of the treatment group for routine problems, and the reverse was true for nonroutine problems. Neither difference was significant statistically.
2. Observation was not a factor in problem-solving achievement because the observed group did not differ appreciably from the unobserved group in achievement.
3. The treatment did not have differential effects among high, average, or low groups (within classes) when prior achievement was taken into consideration. However, the average group scored higher than the high group after adjustment.
4. On the attitude toward mathematics scale the control group scored significantly higher than the treatment group, although both groups' scores exhibited a moderately positive attitude toward mathematics.

Based on self-report information from teacher logs, the treatment teachers followed the teaching requests outlined for them at the orientation session prior to the study. The mean number of daily minutes spent on problem solving was 17 and problem solving was covered on 73 percent of the days school was in session. They averaged over ten minutes per school day on problem solving. Treatment teachers reacted favorably to the project although some had reservations as to its potential for widespread acceptance due to increased preparation time and development of materials.

Discussion

Due to the small sample size of 16 teachers, the results were not expected to reach statistical significance. However, the fact that the control group surpassed the treatment group even after adjustment for initial differences was most unexpected. Several plausible considerations suggest themselves.

Previous research supports the use of instructional strategies as a means to increase problem-solving achievement. The present study was designed to make use of regular classroom teachers in a natural school setting, with teachers using available resources. Since teachers were responsible for generating their own resources, some additional preparation was required. For some this may have become a burden. It may be unreasonable to expect teachers to carry out this type of program without additional feedback or material.

Another consideration which clouded interpretation of the results was that the treatment and control groups were not identical on the pretest measures. Control classes exceeded treatment classes by two-thirds of a standard deviation in prior problem-solving achievement and by one standard deviation on knowledge of mathematics concepts. Although adjustment for this difference was made in the statistical work by using analysis of covariance, this does not imply that the groups were comparable. The fact that teachers were randomly assigned to experimental conditions guaranteed a bias-free assignment, not equivalent groups. The treatment group, being lower than the control, may have had a more difficult time increasing problem-solving achievement than would a group equivalent to the control.

A final rationale for the lack of anticipated results comes from a comment made by a teacher during an interview. He mentioned that he thought the program was beneficial but didn't think it would show up on a test due to the short length of the treatment. He indicated that to bring about desirable results would require, in his opinion, a full year of exposure to the program.

Another surprising result was the significant difference on attitude toward mathematics, with the control group scoring higher than the treatment group. This result is based solely on posttesting, no preexperimental attitude measure was administered. Both groups scored between 3 and

4 on a 5-point scale, indicating a moderately positive attitude toward mathematics. Though the difference was statistically significant, the educational significance is negligible, given that the group means differed by less than 6 points out of a possible 130.

It is worth mentioning that in their interviews teachers indicated a favorable reaction to the program. Some noticed changes in their students' reaction to word problems. Several teachers said they thought student attitude toward verbal problems improved and students were not as apprehensive about working verbal problems as they had been.

Implications

Systematic instruction in verbal problem solving is not a sufficient condition to increase problem-solving achievement in a classroom setting when regular teachers use normally available classroom materials. Additional training sessions with teachers may be appropriate and observational feedback may be helpful in keeping teachers on task. A longer treatment period (a full school year) would be desirable. Finally, in order to carry out this type of program perhaps more materials like problem sets should be made available, because this would ease the preparation burden.

We have concluded from our research experience that the use of the verbal problem-solving manual alone is not sufficient to improve student performance. If used by teachers who are active teachers (model problem-solving behavior, focus on meaning), we suspect that the verbal problem-solving component per se would have a positive impact. However, we know (from our naturalistic research) that many teachers are not active and, hence, we feel that such teachers would need the information and training provided in the general manual before they could benefit from ideas and strategies presented in the verbal problem-solving manual.

Problem Solving in the 1980s

In the 1980s much attention in mathematics education will be given to problem solving. The National Council of Teachers of Mathematics, in *An Agenda for Action: Recommendations for School Mathematics in the 1980s*, advocates in its first recommendation that problem solving become the focus of school mathematics programs. Reports and suggestions from other major organizations, committees, and individuals echo this sentiment.

The net effect that this interest will have on pupil problem-solving performance remains to be seen. Unfortunately, the history of educational movements suggests that a prevailing Zeitgeist often produces much activity but little substantive improvement. The present focus on problem-solving may be an exception to past experience (we hope so), but our experience suggests that if a solid research base is not built in this area, the present instructional interest in improving students' problem-solving skills is unlikely to have any important impact.

One facet of the research stimulated by the problem-solving movement is the systematic review and codification of past research on problem solving (see Lester, 1980, for example). An important distinction that comes from these reviews is that problem solving means quite different things to different people. Some individuals consider the verbal problems found in contemporary textbooks to be legitimate problems for some pupils, others take the view that a true problem will always involve a much more nonroutine situation, and of course, many people fall somewhere between the extremes of this continuum.

The Missouri Mathematics Program did take account of problem solving in designing instructional treatments and in developing appropriate dependent measures. As noted earlier in the chapter, the problem-solving components of the instructional treatment were primarily based on previous research on verbal problems. This approach was reasonable since most previous problem-solving research had been done on verbal problem solving and textbooks gave prominent attention to this type of problem (changing the curriculum was beyond the scope of our work). As was previously mentioned we feel that the verbal problem-solving component of our active teaching model has potential when used in conjunction with an instructional program that has a heavy development component (see Chapter 7). However, we feel that the problem-solving component we developed could be improved if it were adjusted to reflect broader ways of defining problem-solving activity as suggested by Grouws and Thomas (1981).

Research oriented toward developing theoretical models of problem-solving activity may also be quite helpful in designing specific instructional treatments that carefully foster and develop pupil problem-solving ability. The work by Kulm and Bussmann (1980), which considers the abilities necessary at each of eight problem-solving stages, could profitably be used in this way. Similarly, models based on general problem-solving theory, from suggestions on problem solving (Polya, 1957), or from information processing concepts (Wickelgren, 1974), merit consideration when educators design instructional strategies for problem solving. Similarly, clinical research such as that being done by Clement (1982) and Schoenfeld (1982), where student thinking in the act of problem solving is explored in depth, may eventually lead to implications for teacher behavior which facilitates the development of problem-solving skills.

An approach to developing more knowledge about effective problem-solving strategies is through the careful study of teachers known to be effective in this area. This would involve observation of teachers who consistently generated large student gains on a variety of problem-solving measures (e.g., as we did in Chapter 2 for general achievement) and contrasting the resulting profile with a profile of teachers who are relatively ineffective in generating problem-solving gains. These correlational results could then be empirically examined to check their validity (as we did in Chapter 4). Hopefully, research on instruction in mathematical problem solving will flourish along with the problem-solving movement. Without such research the advocacy of "increased attention to problem solving" will have little impact.

Conclusion

We were unable to find rich naturalistic descriptions of how teachers actually attempt to teach problem solving or detailed information about how teacher instructional behaviors relate to student problem-solving performance. We think that attempts to find and study teachers who are adept at problem solving are a needed and important activity.

However, we were able to find many interesting techniques for teaching verbal problem-solving strategies in the literature. We synthesized that literature with our own ideas by writing an instructional manual (see Appendix B). We did find that students' verbal problem-solving performance was significantly better in experimental than in control classrooms. Thus, the data suggest that teachers can make a significant difference in this instructional area as well as in computation and general knowledge of mathematics as discussed in Chapter 4.

In follow-up work we have raised some questions about the use of the verbal problem-solving manual without the more general model. It is our belief that the verbal problem-solving manual in general will have more impact when it is used in combination with the general model. We feel that the general manual provides some important skills and techniques (e.g., need to present more careful demonstrations, provide more consistent feedback, emphasize the meaning of concepts being studied) that are needed but not practiced by some teachers.

6

Experimental Work in Junior High Classes

General Background

From 1979 to 1981, with the support of the National Institute of Education, we conducted a large experimental study in secondary mathematics classrooms. In addition to testing the basic program in upper-grade levels, this research also involved the use of teachers as partners in adapting the elementary program for use in secondary classrooms. In this chapter we will describe both the process of adapting the program for use in secondary schools and the empirical findings that came from the study itself.

Considering the positive results of the two earlier experimental studies in elementary classrooms, we were interested in expanding our inquiry to secondary classrooms. Unfortunately, at the time we were writing the proposal, there was comparatively very little process data which described the normative aspects of mathematics teaching in secondary settings. Fortunately, what data did exist were largely consistent with our treatment program. For example, McConnell (1977) reported that the following teacher behaviors correlated with student learning in high school algebra classes: task orientation, clarity, enthusiasm, and frequent teacher talk. These variables were very similar to the teaching behaviors we were testing in elementary schools. Furthermore, our emphasis on the development portion of the lesson also had some empirical support on the secondary level (see, for example, Zahn, 1966).

Our elementary school data were largely consistent with perhaps the most comprehensive source of information related to effective mathematics teaching in junior high schools (Evertson, Anderson, Anderson, and Brophy, 1980). The findings from the Evertson et al. research overlapped with most aspects of our existing elementary school treatment program.

These researchers found that effective instruction in junior high math classes was characterized by high academic orientation, relatively more whole-group instruction, frequent public recitation and discussion, active student involvement, and maintenance of a rapid pace.

Their results are also consistent with our program and findings in other ways. Both sets of data agree that successful mathematics teachers are more *active* in both public (development/discussion, recitation) and private (seatwork) parts of the lesson. Furthermore, both research programs illustrate that appropriate uses of monitoring and accountability are associated with student achievement, and both suggest that the relationship between lesson parts is critical (e.g., the amount of seatwork time is less important than how well students are prepared for it).

Even though available data in secondary schools were reasonably consistent with the elementary model we had built, we had hoped to obtain a set of naturalistic findings in a specific school setting prior to conducting an experimental project. In the original grant proposal, we had proposed a three-year project comprised of three distinct studies. We would first conduct a treatment study (simply asking mathematics teachers to implement the existing program) and would use the results and the responses of these teachers after they had utilized the program to build a modified and perhaps more sensitive mathematics program for use in secondary settings.

In the second study, with the assistance of a new sample of teachers, we had planned to actively involve teachers in the modification of the training program. That is, the new secondary teachers would have been provided with program material and all related findings, including the results of the first secondary study, and comments by previous teachers who had used the program in fourth-, sixth-, and eighth-grade classrooms. In the third year of the study, we had proposed a retention study in order to determine the extent to which the treatment (of the year before) influenced students in their subsequent learning in mathematics.

Unfortunately, the research program reported here had to be modified to be conducted over a period of 18 months. Several important decisions thus had to be made and made rather quickly. Some of these decisions are described, because they may help other researchers who will be working with practitioners to think through potential problems.

Because of the time limit, it was possible to conduct only one major treatment study. Since we were committed to the idea of involving teachers in reviewing and planning the research, the question became, Under what circumstances could we do this? We had two alternatives: we could work with the teachers in a very quick manner and be able to begin the treatment in the fall, or we could use the fall semester as a way to become acquainted and work with teachers and conduct the experiment in the spring semester.

We chose the former, less involved, partnership arrangement with teachers in order to begin the experiment in the early fall. We thought that secondary teachers might be less responsive than elementary school teachers to research (as described in the popular literature), and that once routines (and plans) were established in the school year, both secondary teachers and students would be more hesitant about changing classroom procedures. Clearly, this assumption in itself is an empirical question and one which merits investigation.

In retrospect, we are not too disappointed with this decision. There are many ways in which a partnership model (working with teachers) can be tested and implemented, and the context of our study dictated that we use a minimum model, with relatively little time for joint decision making and planning. Whether the results would be different under other conditions and with different samples again is a topic for future research.

After having made the decision to begin the program relatively early in the year, there were still many procedural options which were available to us. For example, we could have had three different meetings, reviewing one-third of the training program at each meeting. The teachers could try the program for two weeks and then come back for major consolidation meetings, with the option of revising large parts of the program. We could have paid for substitutes so that some of the partner teachers could observe one another and use this information for modifying the program. We want to emphasize here that external constraints were instrumental in only one decision we made concerning the design of the study. Other options we chose were largely our own, and, in retrospect, even the original constraint does not seem especially important because of the large number of researcher-teacher partnership arrangements which need to be tested.

After deciding to test a treatment program relatively early in the year, we chose to use a minimal partnership arrangement wherein the time for involvement between teachers and researchers was relatively limited, but the decision-making process was still open and all aspects of the program were subject to change. We were prepared to spend as much time with teachers as necessary in order to develop a program which all participants were willing to implement. We ultimately excluded from consideration models which would provide continuing feedback to teachers (the chance to observe or to be observed by fellow teachers) or the opportunity to modify the program in major ways once it had been initiated.

We were very interested in secondary teachers' opinions of the program and the types and extent of modifications they would suggest. In previous interview work with teachers, one of us had discovered that teachers often have interesting explanations for their classroom behavior, even though they appear to be unaware of certain aspects of their behavior (Brophy and Good, 1974). We also had been favorably impressed by the work that Bill Tikunoff, Betty Ward, Gary Griffin, and others (working at the Far West Laboratory) had been doing in building partnership relations with teachers, and had seen in draft form some of the interesting work that had been produced by teachers (Behnke et al., 1981). Although we wanted to work with teachers to modify a program rather than provide resources for teachers to do their own research work, we were encouraged by the potential benefits of involving teachers in program change.

We were also specifically interested in learning how secondary teachers would react to the program, because the recommended instructional techniques had resulted largely from our observation of elementary school teachers. We wanted to learn from teachers whether certain aspects of the program might be inoperative in secondary classrooms.

Prior to meeting with the teachers, we talked with Drs. Carolyn Evertson (then at the University of Texas) and Perry Lanier (Michigan State University) so that we could include their insights in potential modifications of the program. Both researchers were conducting large-scale studies with secondary mathematics teachers and we wanted to take advantage of their research experience. We did not want to provide this information to teachers in advance of our meeting because it could bias their initial impressions and reactions to the program (we didn't want to overload them with experts' opinions). However, we did intend to use Evertson's and Lanier's recommendations at the end of our decision-making conference with teachers, if necessary ("Here is what some other people have said about the program, what do you think of the potential value of their comments?"). As it turned out, both researchers thought the program would be perceived by junior high teachers as similar to what they were already doing and most of their suggestions were related to modifying the program to fit a more mature and responsible secondary-school student population. Several of their comments about modifying the program (e.g., the need for systematizing evaluation standards) were not utilized because of wide variation in teacher opinion. Most of the suggestions which Drs. Evertson and Lanier made were also made by classroom teachers during our meetings with them (including the need for common evaluative practices).

The Planning Meeting with Teachers

Six teachers were randomly selected to be partnership teachers from the volunteer sample of teachers who were willing to participate in the project (the sample will be described below). Prior to attending the meeting, teachers were given both the general treatment manual and the verbal problem-solving manual and were asked to read and critique both. In addition, teachers had an evaluation sheet to fill out and bring to the meeting. We wanted to determine what each teacher thought about the program prior to discussion. From our knowledge of group-discussion literature, it seemed probable that some attitudes of individual teachers might be affected by the particular teachers who happened to speak first or the intensity of the presentations of individual teachers. The raw response forms that teachers brought to the meeting are presented in Appendix I. As can be seen from examining these protocols, the teachers were basically supportive of the program and were willing to attempt to implement it when the session started.

At the beginning of the meeting, each of the principal investigators made a brief five-minute presentation about the scope of the project. Some of the comments by one investigator follow:

> At this point, what we are doing is turning to teachers, the experts in secondary education, because this program was developed for elementary school usage. We're very interested in your criticisms, problems that might develop when the program is used in secondary schools, and we're willing to adapt it as necessary. This meeting is an open invitation for you to react to the program and to accept what appears to be useful and to revise parts that need to be changed. If it seems essentially workable and testable in its present form, that too is okay.

In our remarks we attempted to point out to partner teachers that though the program had worked reasonably well in elementary schools, we had no data base in secondary schools. In essence, we wanted teachers to know that we were very interested in making any and all modifications necessary to make the program work well in secondary schools, and that we encouraged their participation and especially their criticisms of the program.

After we had made our remarks, one of the participating teachers immediately asked the mathematics supervisor to react to the program and to the meeting generally. Among other things, the supervisor said:

> To be clear, you're working directly with the University of Missouri and the school district doesn't control any of the factors of the research. When I first read the program, my evaluation was that the components were important to what junior high teachers should be doing. Although teachers may not sequence all parts of the program like they are here, it seems similar to what many junior high teachers are already doing. I was kind of pleased that we went into the research, because it's hard to get involved in research where there are teacher variables and attempts to improve instruction. In general, I appreciate what they're trying to do and I think that the program is a very good one.

The mathematics supervisor was very positive (perhaps too positive) about the potential benefits of the program. We had requested a meeting with the teachers alone; however, the supervisor also came. He had planned to be an observer only, and his responses then (and a few later in the meeting) were at the request of particular teachers.

What effect his presence had (if any) cannot be determined. However, in subsequent debriefing interviews at the end of the project, two of the partnership teachers were explicitly asked about the presence of the supervisor in independent sessions. These two teachers indicated that they viewed (and that they thought others felt the same way) the secondary mathematics coordinator more as a friendly consultant who was interested in helping and working with them. They did not view him as an administrator whose role was to control teachers. Still, his presence and his expression of positive affect may have reduced some criticisms that teachers wanted to make. However, such criticisms were not given on the sheets that teachers brought to the meeting and there is thus no evidence to indicate that the supervisor had any effect upon the proceedings.

In some ways, his presence may have been useful. For example, when the meeting was turned over to teachers for their input, the very first question was, "How were we selected?" Another participant asked, "I know we're here, but how many schools are involved? Are all schools in the city involved?" The supervisor pointed out that all junior high schools in the city had an opportunity to participate in the project, except for some of the Title I schools, which were involved in another experimental program on math skills. The district did not want the two research programs interfering with one another. He then stated that the investigators had selected the six teachers randomly from a list of volunteer schools.

At this point, one of the participating teachers indicated that the supervisor's account was consistent with her experience. She further noted, "Well, the day that our principal talked to us, he made participation optional on our part. We all kind of sat there and looked at each other and finally I said, 'Do we need unanimous participation in our building?' and he said, 'Preferably.' You know, by that time, we were all saying, 'Why not?'" Our interview data reveal that at least one teacher felt coerced into participation.

Group Discussion

In general, the first few moments of the group discussion were spent talking about general procedural events (Why were we selected?) and the history of the project. Following introductory remarks by the principal investigators and the mathematics coordinator, and general procedural discussions, individual teachers were asked to characterize the strengths and weaknesses of the program from their individual perspectives and to suggest changes that the group might want to consider in modifying the program. To minimize premature evaluations, we asked each teacher to present his or her reactions before we requested general comments and reactions to the ideas which were presented. (We did this in part because at an earlier meeting with secondary teachers, we found that some teachers dominated the discussion—more on this later.) However, it was difficult to adhere to this procedure and we were engaged in group conversation about the wisdom of certain strategies before all teachers had made their comments. Nevertheless, all teachers in the project did have a chance to present their critiques of the program before serious negotiation began about aspects of the program which would be changed.

After each teacher's presentation, group discussion followed, more or less moderated by one of the investigators. Essentially, the dialogue proceeded in the following manner, "Here's a suggestion for change; what do you think about it; what are the advantages and disadvantages?" The decision-making discussion seemed to center on two general topics associated with the project, but not explicitly addressed in project materials. One topic involved the role of testing in the program; some teachers wanted testing to occur on a controlled schedule and other teachers felt that different topics and different types of classrooms necessitated different types and frequencies of testing. Some teachers were adamant that homework had to be graded, and other teachers felt strongly that they did not have sufficient time to grade homework assignments (i.e., they preferred a checking system rather than a grading system). In the end, no changes were made in testing or the grading of homework.

The teachers strongly felt that more time should be provided for review in the junior high setting. They thought that students needed to be actively involved in the review process rather than that review should be conducted as a totally teacher-dominated activity. Accordingly, we decided to extend the length of time allocated for review. Teachers also wanted to incorporate the verbal problem strategies into the program in a very systematic way. They emphasized that verbal problem-solving instruction should take place every day for ten minutes, and that just having students work a few problems at their desks was an insufficient strategy. One teacher pointed out, and others agreed, that teachers need to prepare in order to follow this program and that the verbal problems to be solved should be determined before each class period begins. Verbal problems selected must be carefully interfaced with the development lesson which follows. Another important change which resulted from the teacher meeting was that the time recommended for homework was doubled. In elementary schools the recommended homework assignment took 15 minutes; here, the time increased to 30 minutes. We decided that assignments should be flexible enough to allow for varied rates at which individual students work. Teachers could allow some students to do a portion of their homework in the classroom, but all students would have to do at least some of their homework outside class. Another program change was that students could now be assigned a weekly quiz as part of their broad review time on Monday.

In general, the discussion produced many shared assumptions about what the program was and was not, even though some of the discussion did not lead to changes in the program. The modified aspects of the program and those parts of the program receiving most comment and discussion during the group meetings were summarized by one of the principal investigators and subsequently given to teachers for their approval and/or suggestions. The modifications of the program appear as follows: (1) Teachers' Manual Addendum for Junior High Work; and (2) A New Procedural Summary for the Verbal Problem-solving Manual.

TEACHERS' MANUAL ADDENDUM FOR JUNIOR HIGH WORK

This addendum describes modifications for using the Missouri Mathematics Effectiveness Project Teacher Manual in junior high school classes. The modifications include substantive changes as well as minor adjustments. The changes resulted from a group meeting of junior high mathematics teachers who read the materials and then met to discuss the program. The following revisions reflect the collective thinking of the group.

During the introductory phase of the lesson, a number of things must take place: a brief review, checking homework, and some mental computation. There was agreement that teachers should move rapidly through this phase, because there is a tendency to spend too much time going over homework, and also because the review at this point is distinct from prerequisite skills in the development portion of the lesson. However, in light of the additional math time available at the junior high school level, it was decided to extend the time allotted for the introductory phase to twelve minutes, as shown on the appended Time and Summary Sheet.

The ten minutes following the introductory activities are designated for instruction in verbal problem solving, using strategies outlined in the Verbal Problem Manual. The group endorsed instruction in verbal problem solving, and several teachers pointed out that the ability to compute in isolation is of very little value unless students can use the skills in a variety of situations. It should be emphasized that the time devoted to verbal problem solving should involve the teacher teaching and the students participating in class discussion. This recommendation is *not* fulfilled by just having the students work a few word problems at their desks. It was pointed out by one teacher that teachers need to be prepared to teach in order to follow this program and that this means having the verbal problems to be used in this part of the lesson determined before the period begins. It was also mentioned that there are often a variety of textbooks available at school and these can be a very useful source of problems for this part of the lesson.

The importance of active, careful, and meaningful teaching of the topic for the day in the development portion of the lesson was affirmed. The necessity for a smooth transition from the development phase to the seatwork phase (where students work individually on a collection of problems or exercises at their desks) was mentioned, along with the comment that much time could be lost if this transition is not carefully managed. The program suggests that teachers control practice in the latter part of the development phase; that is, teachers should have students work a problem like the first problem in their assignment and then discuss and demonstrate its solution in front of the class. This procedure should be repeated several times until all the students seem to have the idea. This controlled practice helps a great deal in getting students started immediately after the seatwork assignment is given.

The procedure for seatwork as described in the teacher's manual was not changed. Teachers are to make sure that each student is working before providing individual help. During seatwork, the teacher should circulate about the room, supervising student work to assure that students are not practicing incorrect methods.

There was considerable discussion about how to hold students accountable at the end of the period. It was agreed that this could be done in several ways, including: (1) calling on some students to give their answers to particular problems; (2) checking the answers to the first few problems orally (and then occasionally taking grades); (3) calling on students and asking how many problems they had finished. Other methods are also possible. The important thing is that students do some problems in class while the skills and ideas are fresh in their minds and while help is available for those who need it.

There was general agreement that a homework assignment on Monday through Thursday was a reasonable requirement and that students this age could be expected to work on homework for longer periods of time than those specified in the manual. It was decided that assignments should take 15–30 minutes. This flexibility allows for the different rates at which individual students work and also permits teachers who like to give a combined seatwork-homework assignment to do so. That is, a given number of problems are assigned and students are allowed to work on them during the 10–20 minutes of seatwork in class; whatever problems are not finished are completed outside of class as homework. Under these circumstances, the assignment should be long enough to provide some homework for everyone.

Several teachers mentioned that junior high students need to assume more responsibility than they have previously, and that teachers should thus expect homework to be done on time. Some teachers suggested different ways of recording homework grades (e.g., on some days, scoring the papers and recording the grades, and on other days just checking off complete papers. In this way students would not know which procedure was to be used on a particular day). The way teachers handled collected homework was left to the discretion of individual teachers.

The discussion of the structured reviews every Monday centered on how they should be conducted and the flexibility in scheduling them. There was agreement that the review need not be solely lecture, but should include student interaction in the form of student questions, having students go to the board, or having students work a review problem at their desks with discussion following. The possibility of a weekly quiz during *part* of the review time was discussed and this was deemed acceptable. It was decided that teachers could have this flexibility in scheduling reviews, but when a review was not conducted on Monday, the day it was conducted should be recorded in the log.

The timing of quizzes and tests was discussed and this scheduling was left to the individual teacher, with the understanding that tests and quizzes would be noted in the daily log.

Finally, the topic of occasional variation from the time framework was discussed, and the consensus reached was that teachers should adhere as closely as possible to the guidelines in order to give the program a valid test, realizing that some variance from the guidelines may be unavoidable. However, in such cases, each teacher should make a concerted effort to adhere as closely as possible to the guidelines.

A NEW PROCEDURAL SUMMARY FOR THE VERBAL PROBLEM-SOLVING MANUAL

The ability to solve verbal problems, or word problems, is an important skill. In order to insure that verbal problem solving will receive systematic attention, each teacher is asked to spend ten minutes of every mathematics class period on this topic. Students are thus exposed to verbal problem solving daily, even when problem solving is not the principal objective of the lesson. This ten-minute period should be a time when the teacher teaches problem solving and *not* a time for students to sit passively in their seats and work a couple of verbal problems. That is, the teacher should actively model the solving process, have students work a problem, and then have a class discussion of ways to solve the problem, and so on.

Preparing to actively teach verbal problem solving is not always easy. To assist in the preparation of this part of each lesson, a *Verbal Problem-solving Manual* has been developed. It details five instructional strategies which seem to be associated with improved student performance in this area. We ask that you use one of these strategies each day.

The strategies in the manual can be used independently or in combination. Some will be more appropriate for particular problems or topics than others, and the choice of which strategy to use on a given day is left to the discretion of each teacher. Since each of the ideas has merit, and since there is value in using a variety of ideas, it is important to use each idea regularly. Whether one idea is used for a week at a time, or a different idea is used each day is not important. To insure that every idea gets some exposure, we ask that each idea be implemented at least once every two weeks.

The procedures outlined in the *Verbal Problem-solving Manual* are designed to be used with textbook problems and your textbook will be the primary resource for these problems. Other resources can include textbooks no longer in use, textbooks from lower grade levels, and problems based on information generated during class discussions, field trips, and so on.

A table summarizing the key points of each idea in the manual is attached to aid you in viewing the program at a glance.

The *Verbal Problem-solving Manual* is the detailed resource to assist you in the application of the ideas.

Each teacher is asked to keep a record of the verbal problem-solving activities used. The log should include the amount of time spent on problem solving, the strategies used, the in-class and homework assignments on problem solving, and any exceptions or conflicts which affect the program. By exceptions we mean situations which arise from time to time and affect your schedule: shortened periods, cancelled classes, holidays, testing days, and so on. Please note these occurrences in the log, as well as any other conflicts. The logs will be collected every two weeks in order that progress and treatment implementation can be monitored.

SUMMARY OF VERBAL PROBLEM-SOLVING PROGRAM

Time: Ten minutes per day, every day
Techniques:

Problems without Numbers

1. Recast textbook problems so that no numbers appear.
2. Prepare these *ahead of class time* on an overhead transparency, worksheet, chalkboard, etc.
3. Focus on how to solve each problem.

Writing Verbal Problems

1. Use graphs, charts, tables from the textbook, newspapers, etc., and have students formulate problems based on these data.
2. Use data from situations that arise (field trips, sports, etc.).
3. Have students supply their own data.
4. Have students solve each others' problems.

Estimating Answers

1. Show students *how* to estimate.
2. Have students estimate orally.
3. Estimate answers to text problems before doing computation.
4. Eventually have students estimate answers to all verbal problems they work (underline estimate, circle exact answers).

Reading Verbal Problems

1. Focus on word recognition, context, general comprehension.
2. Write, pronounce, define new words; give examples and non-examples of the concept.
3. Read problems aloud; use tape recorders.
4. Provide reading assistance on an individual basis.
5. Have students and teacher alternately read and discuss problems.
6. Use text problems, student-created problems, problems from older textbooks.

Writing Open Sentences

1. Translate conditions into equations.
2. Allow for equivalent equations.
3. Use problems from lower grade levels.

In many ways, it seemed that the discussion of the program led to some increase in teachers' willingness to implement it, but also perhaps to more variance in the program (in terms of individual teachers' interpretations) than was present in the previous elementary school sample. That is, group discussions *at the time* appeared to support the notion that the recommended times for each part of the lesson were generalized statements about an average distribution of time over the year, and that individual lessons might deviate from the average. Although the investigators occasionally mentioned that balance among the lesson parts was important, the need for adjustment from lesson to lesson was also frequently expressed. We suspected that teachers in this sample would be much more likely to modify the time allotments than elementary school teachers and secondary teachers who subsequently were asked to use the treatment but did not participate in the modification process. (However, interview information collected subsequently suggests that the tight time lines presented in the manual and perhaps information given at the training session led many teachers to feel that they should not vary time lines.)

Although there were not many alterations in the program, we feel that the changes made were useful and important. In addition to their substantive contributions, teachers who were involved in the discussion may have been very important symbolically to subsequent work with teachers who were asked to use the program. That is, the knowledge that other secondary teachers had examined the program may have been instrumental in obtaining teachers' cooperation and participation and perhaps increased their adherence to program suggestions.

Outside Evaluation

The meeting with the partnership teachers was tape-recorded in order to allow for outside evaluation about the conditions and processes of that meeting. We were very fortunate to have the professional consulting services of Dr. Dee Ann Spencer Hall, a sociologist trained in the qualitative tradition (symbolic interaction) and a former classroom teacher. She is very interested in classroom process and has completed several field studies in classrooms. Although she is a colleague and friend, we were confident that Dr. Spencer Hall would provide a rigorous critique of the research and bring to the study a perspective which complemented but broadened the perspective of the principal investigators (we are pleased to report that our positive expectations were fulfilled).

Dr. Spencer Hall listened to the tape and concluded that teachers had an opportunity to contribute to the modification of the program and were encouraged to do so. However, she indicated that teachers appeared to be stating beliefs and suggestions which tended to present and support their own teaching practices rather than proactively dealing with the program as a means of developing an approach to mathematics instruction. These were veteran teachers who in some ways seemed to be justifying their current teaching practices more than they were searching for new and integrative approaches. At a minimum, Dr. Spencer Hall felt that the program did develop some common boundaries, set the conditions for minimum participation, and produced a few program modifications (although this varied from teacher to teacher). In subsequent interviews with the teachers in the program (after the program had ended), she found that many teachers were concerned about the number of changes which were occurring in the district (for example, the closing of several schools and the shifting of teachers from one building to another), and she suggested that many of these factors may have led teachers to be more personally oriented than program oriented during the decision-making meeting.

Context Effects

We were sensitive to possible context effects in the data. Another group of teachers might have different reactions to the program and their own unique suggestions for modifying the program. This is possible for a variety of reasons, including the individual teachers attending the meeting and the nature of the district and its relationship to the community. Although we did not have the time or resources to examine these context effects directly, we did want to examine them indirectly. We wondered if teachers who were making recommendations about a program they would implement would evaluate the program differently from teachers who did not have to use the modified program. One could argue that teachers who would soon teach the program would become much more interested in it and thoughtful about it. On the other hand, it could be that teachers who do not have to utilize a program (and perhaps have to do extra work) may feel freer to make more recommendations than they would if they had to implement those changes in their own instructional programs.

To consider these context effects, we met in a different city with another group of junior high teachers who knew that they would not be required to use the program. (This meeting took place prior to the meeting with users—the partnership meeting described above.) These teachers were also given both training manuals to read prior to the meeting and were requested to bring their critiques of the program when they attended the meeting. The teacher responses indicate that these teachers were very supportive of the program.

After these materials were collected, there was a discussion with the teachers similar to the one reported above. Teachers were asked to help us improve the program and were encouraged to make criticisms and suggestions which would make the program more practical for use in secondary classrooms. These teachers expressed much more positive affect about the program during the meeting than did teachers who would subsequently use the program. In general, the questionnaire reaction of both user and nonuser teachers to the treatment program was very favorable.

Dr. Spencer Hall attended this meeting with nonusers as an observer and her reactions to this group of teachers were very similar to our own. She noted that nonusers were much more positive in describing the program and showed more interest in its potential value for teaching mathematics. She also noted that these teachers were much more energetic, younger, and more knowledgeable (at least more expansive) in their discussion of mathematical concepts and instructional strategies. In general, both the principal investigators and the outside consultant/evaluator felt that this group of nonusers would have been much more enthusiastic in their implementation of the program than the other group. However, it is impossible to determine whether our beliefs would have been matched by actual behavior. At a minimum, these data suggest that the modifications suggested by teachers were reasonably consistent across two different samples and that the condition of teaching or not teaching the program did not appear to mediate suggestions (although affect and involvement did vary between the two groups).

As noted above, Dr. Spencer Hall had access to the tape recording of the user meeting, had observed the nonuser meeting, and subsequently had the opportunity to interview several of the user teachers. More of her comparative comments about the two samples can be found elsewhere (Good and Grouws, 1981); however, it is interesting to note that she felt nonuser teachers were much more receptive to the program than were users. What effect, if any, such apparent differences in teacher attitudes would have on program implementation and impact is, of course, impossible to determine. However, such data do suggest that the context of the research site in teacher partnership work may be very important.

METHOD

Sample

The research took place in a large southwestern city. The investigators met with school administrators during the summer in order to explain the project and to obtain permission to do the study. School administrators explained the project to principals, who in turn described the project to classroom teachers. All junior high schools in the district were contacted, except for several Title I schools which were participating in another mathematics experiment.

In some schools all eighth-grade teachers volunteered for the project; in other schools, only one teacher volunteered. After determining the number of teachers who would be participating in each school, we again discussed the sample with the school administration. We wanted to balance the sample as closely as possible according to the schools that control and treatment classes came from. Because we had to meet with partner teachers early in the year, it was impossible to derive the sample after the administration of the pretest (the best research strategy). Based upon information the school district gave us about the student population each school served, we divided the sample into matched sets of schools, and randomly assigned schools to treatment conditions.

Nineteen teachers volunteered for the project. Of these, six were assigned to the partnership group and five teachers were assigned to the treatment group. Both the partnership teachers and the treatment teachers were asked to use the instructional program in their classrooms. The only difference between the two groups was that the partnership group had a chance to modify the program, and the treatment group did not.

All control teachers allowed us to observe their classrooms, and to collect pre- and postachievement data in their classrooms. However, three of these teachers did not attend the orientation training (we called these teachers nonparticipating controls to distinguish them from the control teachers who received a motivational treatment).

The 19 teachers were drawn from 12 different junior high schools. Most of the target classrooms in the study were regular eighth-grade classrooms. However, because one partnership teacher who was basically teaching algebra suggested at the partnership meeting that it would be interesting to compare algebra classrooms as well, we did build a minor, pilot algebra study into the design. The distribution of regular eighth-grade classes and eighth- and ninth-grade algebra classes across the entire sample was as follows: partnership teachers, ten, one, and three; treatment group teachers, seven, zero, and two; regular control, nine, three, and one; nonparticipating control, five, zero, and zero. The final sample is summarized in Table 6.1.

Teacher Training

Two weeks after the initial meeting with the partnership teachers, the project directors met with 16 of the teachers from the 12 schools, the school principals, and district administrators. At the training workshop all participants were told that the program was largely based upon earlier observation of relatively effective and ineffective fourth-grade mathematics teachers. Furthermore, we explained that subsequent research in fourth- and sixth-grade classrooms had provided experimental data to illustrate that students in classrooms of teachers who had been exposed to the treatment did better in some areas of achievement than did students

TABLE 6.1   Teacher/Class Sample: Junior High Mathematics Study

Partnership Teachers	
# Teachers	6
# Regular eighth-grade math	10
# Eighth-grade algebra	1
# Ninth-grade algebra	3
Treatment Group	
# Teachers	5
# Regular eighth-grade math	7
# Eighth-grade algebra	0
# Ninth-grade algebra	2
Control Group—Observation	
# Teachers	5
# Regular eighth-grade math	9
# Eighth-grade algebra	3
# Ninth-grade algebra	1
Control Group—No Observation	
# Teachers	3
# Regular eighth-grade math	5
# Eighth-grade algebra	0
# Ninth-grade algebra	0
Combined Partner and Treatment	
# Teachers	11
# Regular eighth-grade math	17
# Eighth-grade algebra	1
# Ninth-grade algebra	5
Combined Control—Observation and No Observation	
# Teachers	8
# Regular eighth-grade math	14
# Eighth-grade algebra	3
# Ninth-grade algebra	1

in control classrooms. We also told participants that a group of junior high teachers from their own district had been working with us to modify and hopefully to improve the program. Teachers were informed that they were going to be requested to teach the modified program.

Teachers were told that although we expected the program to work, the earlier correlational/experimental work had been conducted in elementary schools, and the present project was the first test of the program in secondary schools. After a brief introduction, the teachers and their principals were divided into two groups. Teachers in the treatment group (including both partnership and regular treatment teachers) were given an explanation of the program (the training lasted 90 minutes). After the training session, regular treatment teachers received the basic manual (see Chapter 3) along with instructions to read it and to begin to plan for implementation (partnership teachers already had read the manual). In this manual, definitions and rationales were presented for each part of the lesson, with detailed descriptions of how to implement the teaching ideas. In addition, treatment teachers were also told what modifications were made by the teachers at the planning meeting and received the verbal problem-solving manual and the more precise procedural directions which had been developed by teachers at the partnership meeting.

Control teachers were told that they would not receive details of the instructional program until later in the year, but that we hoped this information might be especially useful to them then. At that time they would receive information (i.e., other teachers' comments) about the program itself. Finally, we informed control teachers that their immediate role in the project was to continue to instruct in their own styles. Because the control teachers knew that the research was designed to improve student achievement, the school district was interested in the research, and they were being observed, we feel reasonably confident that a strong Hawthorne control was created (as noted previously, three control teachers did not attend this orientation meeting).

Observations

Control and treatment teachers were observed approximately 12 times during the study. Each classroom in the project was observed four to seven times, depending on the number of classes the teacher had in the study. If the teacher had three or four classes, then only four or five observations were made in each of the teacher's classes. If the teacher had only one or two classes involved in the study, then each class was observed six times. Each observer conducted approximately one-half of the observations in each of the classes and observations in each class were equally spaced throughout the duration of the study.

All observations were made by two doctoral students in mathematics education who were living in the target city during the project. The observers were trained initially with written transcripts and videotapes, and then in actual classrooms. Observers reached reliabilities of .80 on each of the coding distinctions used in the actual study.

Schedule for Meetings and Testing

The initial meeting with partnership teachers took place in late September and the training/briefing session with all the teachers was held during the first week in October. Pretests were administered in the second week of October and classroom observations began shortly thereafter. The posttest was administered in January; hence, the treatment lasted about three months. The mathematics pretest had been used in the Texas Effectiveness Project and was provided to us by Dr. Evertson. The post-achievement tests were two subtests of the SRA, Level F, form 1 test (Math Concepts and Problem Solving). Reliabilities on both instruments are excellent.

Program Implementation

Teacher Opinions. After terminating the project, we wanted to obtain teachers' perceptions of their involvement in the program. We felt that it would be useful to have someone else collect these data for us, and as reported earlier, Dr. Spencer Hall was willing to do this. She traveled to the target city about a month after the project had been concluded, and interviewed all six of the partnership teachers and three of the five treatment teachers (she also interviewed two of the control teachers, but those data are not relevant to the present discussion).

The interviews lasted about one hour; a few of them had to be conducted via phone (because of an ice storm). Dr. Spencer Hall described herself as a former teacher and a sociologist interested in ideas people have about teaching. She indicated that although she wanted to ask several questions about the project, she was not one of the project staff and simply wanted to know what their reactions were to the program. She stated that criticisms were welcome.

At the conclusion of her interview work, she drafted a brief summary for the project directors. A very condensed version of her comments appears here (for more information, see Good and Grouws, 1981).

> In retrospect, I think the interviews went quite well. Most of the teachers appeared to be open and honest with me in their responses. I plan to give you feedback in two ways. First, I have enclosed the interview schedule, the questionnaire I used (with modifications depending on which category a teacher was in), summaries of each teacher's interview, and my own subjective reaction to the interviews. Second, I am having the tapes transcribed because I feel that many comments were worth having verbatim and in their entirety.
>
> At this point, I would like to make a few subjective comments. My first reaction to the teachers was that they were all *very* experienced. For example, 5 of the 11 teachers have taught 20 years, two have taught over 15, and the remaining 4 have taught over 12 years. The range of 12–20 years, with the mode of 20 years gives you an interesting sample. I realize you will get this data from the questionnaire you had them fill out, but the thing that struck me in the interview was that (a) 3 of the 5 who had taught 20 years were obviously burned out on teaching and (b) 4 of these 5 were sure that kids today have "gone to the dogs." (Although some in the over 15-year range expressed the same attitude.) My opinion is that such dissatisfaction with teaching and the attitude that kids today are disturbed, of low ability, and generally hyper and out of control affects these teachers' effectiveness—and ultimately, in this case, reduced their enthusiasm for trying something new, i.e., your program.

In summary, as I said before, these comments are subjective and only somewhat systematic at this point; they only relate to general areas and concerns. The interviews, however, did reinforce and substantiate my own opinion that the data should be contextualized. Even though the data are primarily quantitative, your methodological strategies have taken you into the qualitative area, through partnership information discussions and my open-ended interviews, and thus more consideration should be taken of the contextual features of day-to-day life in the schools and classrooms. For example, what about the context of a sixth-grade class is different from an eighth-grade class and would make implementation of the program more problematic for eighth-grade teachers? The situational context in a junior high is probably more disruptive, more complex (due to size, for example, and having six or seven different classes), has more conflicts, and is less predictable (in terms of events and individual behaviors).

The interview schedule Dr. Spencer Hall used included the following questions:

I. *School*
   1. What is it like to teach in this school?
   2. What kinds of things *inhibit* what you do in your classrooms (problems which interrupt your work)?

II. *Teaching*
   1. How do you feel about teaching, i.e., how satisfied are you with teaching as an occupation? (Probe)
   2. What things might happen to increase your satisfaction?
   3. What bothers you most/least about teaching?
   4. Do you think teaching math is any different from teaching any other subject? (How? In what ways?)
   5. Do you teach differently in different math classes; in the same course at different hours of the day? (Why? How?)

III. *Math Effectiveness Program*
   1. To what extent do you feel you've become involved in the program?
   2. Did you change anything about the way that you've taught math as compared to past years? (What? How?)
   3. Do you see this as a positive change—is your teaching more (or less) effective now?
   4. What do you see as the major strengths of the program?
   5. Weaknesses? (Why)
   6. If you could change the program in any way(s) what would you do? (Why) Additions? Deletions?
   7. Would you (continue to) use it if changes were made?
   8. Did you communicate with other teachers about the program? (How? What happened?)

IV. *Relationship to Researchers*
   1. Did you feel you had adequate input into the program during your early discussion with the researchers?
   2. Did you feel you had an impact—that your comments and suggestions were taken seriously?

3. What (if anything) could have been done to have made communication with the researchers more effective or helpful?
4. What is your *general* feeling about classroom teachers and researchers working together on instructional projects? (Probe)

V. *Outside School*
 1. What do you do when you're not at school?
 2. How much of your time outside school is spent in planning for your time inside school?
 3. Did participation in this program change your planning time outside school? (How?)

At some point we hope to analyze these interviews to determine teachers' beliefs about general aspects of teaching, and to see if these beliefs can be related in any way to effects on students.

Our own analysis of the teacher interview responses to questions Spencer Hall raised about program implementation is that the responses approximate a bell-shaped curve. One partnership teacher (01) had especially strong negative feelings. This teacher reported that the program had no effect upon her/his behavior and that he/she felt coerced into participating in the experiment. The observers felt that the teacher implemented the program at a minimum level despite these reactions. Another teacher (03) reported compliance with certain aspects of the program, but observers found virtually no involvement and/or participation in the study. Other teacher comments about the program in general were reflected in observational data (to be discussed later). For example, one teacher (02) who liked the program because of the verbal problem-solving strategies did emphasize verbal problem solving in the classroom.

Four partnership teachers reported that they liked the meeting and that the introductory meeting with researchers and the subsequent training session went well. Two of those who approved of the program pointed out specific parts that were good; the other two emphasized the similarity between their own teaching and the present program. Similarly, the treatment group had mixed reactions to the program and to the research training session (they did not participate in the modification session).

Both groups of teachers reported that the program was similar to what they were already doing and indeed, Drs. Lanier and Evertson had mentioned this possibility to us previously, suggesting that it was both an advantage and a disadvantage. It was an advantage in the sense that teachers would be willing to implement the program and to make some modifications in their behavior. However, teachers might not be motivated to look for some of the subtle (but important) differences between the program and their present teaching methods.

One year after the project ended, 15 of the project teachers were interviewed (to be described later). Although teachers were not explicitly asked about their reactions to the treatment program, all treatment teachers initiated on their own some comments about a particular part of the program which was meaningful to them. Some teachers said that though they had been conducting class review sessions in the past, the program had prompted them to think systematically about the nature of review and how they could build broader reviews into the lesson. Other teachers indicated that the program had helped them to consider instruction in problem solving and to emphasize this topic more than they had. Some participants commented that their general approach to development and seatwork assignments was very similar to the approach advocated by the program, but that they were now much more careful to be certain that students were ready for seatwork before they assigned it. In general, teachers did not make comments about the program as a whole; rather, they chose to comment upon particular parts which seemed especially meaningful to them. We suspect that teachers were surprised by the general similarities between the experimental program and techniques they had already been using in the classroom. The contrast between their expectations for the program and the actual program may have made the similarities (and hence their reports of them) highly salient. However, their behavior during the program (as we shall see below) and their comments at the end-of-project interviews suggest that some teachers were noting and responding to subtle differences between their teaching practices and program recommendations, at least in some areas of the program.

Observational Data. An initial step in analyzing data was to determine the extent to which partner and treatment teachers had implemented the program. Because participants reported that the program was similar to instructional methods they were already using, it was important to determine whether treatment teachers and control teachers differed in any systematic ways in their classroom behavior (i.e., was the treatment condition associated with some distinctive teaching behavior?).

Table 6.2 presents selected implementation data for all project teachers. The first six teachers are partnership treatment teachers. Teachers 7 through 11 are regular treatment teachers, who were asked to implement the program and were trained to implement the program, but did not have a chance to modify the program. Teachers 12 through 16 were control teachers who attended the orientation meeting and who were observed. As noted before, there were three other teachers who served as controls, and who were observed but who did not attend the orientation meeting (these control teachers were called nonparticipating).

Our first task was to examine the implementation scores to determine if partner and regular treatment teachers were using the program. From just the variables presented in Table 6.2, it is clear that some teachers implemented the program more fully that did others. The average imple-

TABLE 6.2 Selected Implementation Scores by Junior High Teachers*

Teacher Number	Treatment or Control	Average Number of Minutes on Mental Comp.	Average Number of Minutes on Verbal Problem Solving	Average Number of Minutes on Development	Average Number of Minutes on Practice Seatwork	Development Overall Quality	Average Implementation Score
1	T	2.00	8.86	3.43	5.00	2.50	2.57
2	T	7.13	8.53	8.20	9.27	3.92	3.20
3	T	0.00	0.47	6.07	18.20	2.25	1.20
4	T	3.73	6.47	7.80	9.20	3.82	3.47
5	T	4.94	2.75	13.25	4.69	3.14	2.38
6	T	0.00	2.79	4.43	12.86	2.75	2.21
7	T	1.13	4.53	14.53	16.20	3.23	2.33
8	T	1.25	1.75	18.19	11.00	3.85	2.44
9	T	2.00	2.47	6.40	13.93	3.63	3.07
10	T	0.00	4.67	11.80	11.33	2.13	1.89
11	T	2.00	3.00	11.21	12.14	3.17	1.93
12	C	0.57	3.21	14.00	11.36	3.54	1.93
13	C	0.00	4.57	9.71	16.14	4.57	1.86

14	C	0.00	0.00	5.29	22.50	2.44	1.21
15	C	0.00	0.40	13.80	14.53	2.50	1.80
16	C	0.00	0.00	17.00	11.00	2.57	2.33
17	C	0.00	0.00	6.75	26.81	2.15	1.06
18	C	0.00	1.86	1.71	8.71	3.33	1.43
19	C	0.00	0.77	15.46	18.54	3.73	2.77

* Teachers 1–6 are partnership teachers; 7–11 are treatment teachers; 12–16 are regular control teachers; 17–19 are control teachers who did not attend the orientation meeting.

mentation score is based upon the five variables presented in the table, as well as several other variables (including the assignment and checking of homework, the presence of controlled practice, the presence and quality of review work, etc.).

The information presented in Table 6.2 allows the reader to see the variability *between* and *within* treatment and control teachers on selected aspects of the program.

A general implementation score was assigned to each teacher at the end of each observation. The score assigned to the teacher was based on the following scale:

(5) If the teacher implemented all major components

(4) If the teacher implemented most of the major components

(3) If the teacher implemented about one-half of the program components

(2) If the teacher implemented some of the program components

(1) If the teacher implemented very little of the program

Intercoder reliability was estimated on the implementation scores on the basis of seven dual observations made in the target school district during the first two weeks of the study. Perfect agreement was found on five of the seven observations and only a one-point variation found for each of the other two observations.

At the end of the study, a comprehensive implementation score was determined for each teacher, by averaging the individual implementation scores assigned to the teacher at the end of each visit. These means are reported in Table 6.2.

One important question to raise is, does degree of program implementation correlate with residual gain in computation, problem solving, and attitudes toward mathematics? To answer this question teachers' average implementation scores, average time spent on mental computational instruction activities, and average time spent on verbal problem-solving activities were correlated with students' residual scores for computation, problem solving, and attitude. These results are presented in Table 6.3.

As can be seen in Table 6.3, the average implementation score does not correlate significantly with students' residual performance on the computational or problem-solving test. However, average implementation score was found to correlate significantly with students' attitudes toward mathematics ($p = .02$).

The correlation of instructional time spent on mental computation and instructional time spent on verbal problem-solving activities is also presented in Table 6.3. These parts of the program were computed sepa-

TABLE 6.3 The Correlation and *p*-values of Average Implementation, Time on Mental Computation, and Time on Verbal Problem Solving with Residual Gains in Computation, Problem Solving, and Attitudes toward Mathematics

	Average Implementation	Average Instructional Minutes Spent on Mental Computations	Average Instructional Minutes Spent on Verbal Problem Solving
Computational residual (N = 19)	.16 (NS)	.24 (NS)	.45 (.05)
Problem-solving residual (N = 19)	.26 (NS)	.49 (.05)	.51 (.02)
Attitude residual (N = 16)	.56 (.02)	.34 (NS)	.43 (.09)
Average implementation	.63 (.003)	.63 (.003)	.58 (.008)
Time on mental computation	.63 (.003)	——	.63 (.003)
Time on problem solving	.58 (.008)	.63 (.003)	——

rately because they were presumed to be relatively novel instructional acts (not frequently engaged in during secondary mathematics instruction).

As can be seen in Table 6.3, the average time spent on mental computational activities did correlate significantly with students' problem-solving residual scores ($p = .05$), but *not* with their computational scores nor attitudes toward mathematics. Interestingly, instructional time spent on verbal problem-solving activities did correlate significantly with students' residual scores for computation ($p = .05$), problem solving ($p = .02$), and virtually reached a significant relationship with students' attitudes toward mathematics ($p = .09$).

An examination of Table 6.2 shows that partner and regular teachers were found to implement some aspects of the program more often than did control teachers. The mental computation and verbal problem-solving activities differentiated the two samples most sharply, and there was considerable overlap between treatment and control teachers in terms of the amount of time spent on development and seatwork, and in the overall quality of the development lesson. Overall, these data suggested that somewhat different behaviors were occurring in treatment and control classrooms and, hence, an analysis of achievement effects would be meaningful, especially in the area of problem solving.

After examining the implementation data, we decided to eliminate Teacher 3 from subsequent data analyses because this teacher was generally deficient in implementing the program. In particular, this teacher did not utilize the mental computation activities and virtually ignored verbal problem solving throughout the course of the experiment.

Observer Opinions. At the conclusion of the project we asked the two observers to describe briefly each teacher they had observed and to provide their general impressions of the teachers, their effectiveness, and the extent of their program implementation. In general, their comments reflect important variations among teachers in the extent to which features of the program were present in control and treatment classrooms. Observers' comments also generally support the implementation data derived from actual observational records.

All in all, the implementation data from three perspectives suggest that treatment teachers generally saw the program as quite similar to teaching techniques they had been using previously. However, there is evidence that some treatment teachers were influenced by the program, and that they did instruct in a manner which differed in some ways from control teachers. The data further suggest that teachers were more influenced by certain aspects of the program rather than the treatment as a whole. The verbal problem-solving techniques and mental computational activities appear to have had most impact upon treatment teachers.

Results

After deciding to eliminate Teacher 3 from all analyses dealing with program effects on students (because of this teacher's low implementation score), two general questions affecting the analysis of the data were considered. First, we wanted to know if the algebra teachers in the treatment and control conditions differed in any way (it was assumed that algebra students would do better than nonalgebra students on the posttest). This analysis showed no differences between the algebra teachers in the treatment and control classrooms (the p-value for postcomputation was .71; the p-value for post-problem–solving was .34; and the p-value for post-attitude was .80). Next we wanted to determine whether the partnership teachers differed from regular treatment teachers, and if these groups should be analyzed separately. These comparisons suggested that there were no significant differences between partnership and regular treatment teachers. The respective p-values were: postcomputation, .50; post-problem–solving, $p = .91$; and post-attitude, $p = .42$. The effects of partnership and treatment teachers were *not* different.

As can be seen in Table 6.4, the preachievement level of students in the control group was somewhat higher than achievement levels for treatment classes. Despite this, raw scores of students in the treatment group on both the postcomputation and the post-problem–solving sections of the SRA achievement test were somewhat higher than scores of students in control classrooms. Analysis of covariance procedures (using the pre-scores as a covariate) were performed on the classroom means shown in Table 6.4. In performing these analyses, the classroom was used as the unit of analysis and each class that a control or treatment teacher taught was included as a separate unit in the analysis ($N = 39$). The results of these ANCOVA analyses indicated a weak effect in favor of the treatment group on the computational performance of students ($p = .15$). However, the effect of the treatment on problem-solving scores of students was significant ($p = .03$). These results are consistent with the implementation data reported earlier. That is, there was not much variability in general behavior between treatment and control teachers; however, there were noticeable differences in the use of mental computation activities and verbal problem-solving activities.

Follow-Up Teacher Interviews

One year after the formal project had ended, arrangements were made to interview available classroom teachers. Sixteen of the teachers who had participated in the project were still teaching in the school district, and it was possible to interview 15 of them.

TABLE 6.4 Pre-, Raw, and Adjusted Means for Junior High Treatment and Control Classes on Pre-Math Test and on Post-SRA Subtest Scores

	N	Preachievement	Post $\bar{x}$ Computation	Adjusted Computation	Post $\bar{x}$ Problem Solving	Adjusted $\bar{x}$ Problem Solving
Treatment	21	47.65	29.75	29.84	21.90	21.98
Control	18	48.37	28.97	28.86	20.99	20.83

Dr. Spencer Hall had interviewed many of the teachers when the project concluded; her questions focused upon teachers' reactions to the project and their general beliefs about teaching and schooling. The follow-up interviews dealt specifically with the teaching of mathematics and teachers' beliefs about mathematics teaching. In combination, the two sets of interviews should provide an integrated picture of the beliefs that these secondary teachers had about teaching generally and about mathematics teaching more specifically.

In conducting these interviews, we were fortunate to have the professional consulting services of Dr. Jere Confrey at Mount Holyoke. Dr. Confrey is a former secondary mathematics teacher and is a specialist in mathematics content at the secondary level. She has also done extensive field work involving methods of interviewing students and teachers. She has worked with Dr. Perry Lanier and others at Michigan State University on an intensive study supported by funds from the National Science Foundation. Their research examines students' experiences (and their conceptualizations of mathematics content being taught to them) in junior high general mathematics classes. Dr. Confrey assisted both in the conceptualization of the interviews and in the data collection. Some of the interview questions resulted from the National Science project at Michigan State University. However, many of the questions raised were developed specifically for our own research project.

These interview data were collected in the spring of 1981 and the typed transcripts are now being analyzed. We expect to integrate these analyses with other data sources in the project. For example, Dr. Tom Cooney, a specialist in mathematics education at the University of Georgia, has read certain aspects of the teachers' interviews and has ranked the teachers in terms of their apparent knowledge of mathematics. We are presently collecting rankings from others and at some point we anticipate conducting analyses to see if certain teacher beliefs can be related to classroom behavior.

Appropriateness of Student Achievement Measure

As a result of the teacher interviews it was possible to examine the adequacy of the achievement test from the perspective of classroom teachers. The reliability of the SRA (Level F, Form 1) mathematics test is excellent; however, one can still raise questions about validity. We wondered to what degree the test overlapped with the material teachers taught in class and the types of problems presented in textbooks. We attempted to obtain this information in two different ways.

First, we asked Dr. Confrey to conduct a content analysis to compare the SRA test with the *Holt School Mathematics* text (the book most used in the sample). Her comparisons introduced a number of questions about the adequacy of the test in certain areas (especially general computation). She raises some important concerns about the problem-solving test and the text (the test is a narrower definition of problem solving), but concludes that there is a reasonable congruence between the general text and test (the largest discrepancy occurs between the test and the chapter in the book devoted to problem solving). Her comments raise a number of interesting and important issues about matching text content and test problems.

When the 15 teachers were interviewed a year after the project ended, they were given a copy of the SRA test and were asked to critique each test item (Was it taught? Was the question asked appropriately?). All teachers subsequently returned the test ratings by mail. Their responses generally indicated satisfaction with most of the test items, with only two teachers (one treatment, one control) having strong negative reactions to the test.

Although some problems exist with the match between test and instruction, the test appeared at least minimally adequate for making comparisons. Also, the test appeared to be equally appropriate for treatment and control teachers (as indicated by teacher responses and by our examination of teacher logs).

Treatment Effects: Secondary Level

The data collected in the project indicate that change in teacher behavior and in student performance in secondary schools is possible. In particular, the results demonstrate that participation in the treatment program was associated with a significant positive effect upon students' problem-solving skills, as measured by the SRA test. Although we do not feel that the problems in this test are a completely adequate measure of problem solving, they do represent some skills that appear to be important. It is therefore edifying to see that treatment teachers had a positive effect upon student performance on the problem-solving subtest.

Although we have raised some questions concerning the adequacy of the content criterion test and the level of teacher implementation, the overall evidence suggests that treatment teachers did implement more problem-solving strategies than did control teachers and that the test was a reasonable measure of content being presented in classrooms. We can thus confidently say that the program had a positive effect on treatment teachers' implementation and students' performance on the test.

The data also reflect a moderate, positive trend favoring treatment students' postperformance on the computational subtest of the SRA inventory. However, because the difference is minor, it is probably appropriate to state that the training program had no notable impact upon this type of student achievement. This appears to be the case because both treatment and control teachers taught in similar ways during the developmental portion of their lessons. The qualitative ratings which observers made of all teachers are not high and indicate that much future research needs to focus upon conceptualizing and implementing the development stage of the lesson. Although some teachers in treatment and control classrooms were able to conduct development successfully, most qualitative ratings of implementation were not uniformly high. (See Chapter 9 for more discussion on development.)

The teacher interview data are still being analyzed, and from these data we may be able to make statements about the relationships between teacher beliefs and teaching performance, and their effects upon students. These results will also lead to a fuller understanding of the perceptions that secondary teachers have about teaching mathematics and the conditions under which they teach. In particular, interview data will make us more sensitive to some of the difficulties involved in attempts to change teacher perceptions and behavior and will also make us more aware of the difficult circumstances in which some teachers teach.

Working with Teachers

We began our program of research several years ago by observing what elementary teachers (who were more and less successful in obtaining student achievement) did in the classroom and by building a training program that was sensitive to those differences (although it included a few other components as well). In the present research we attempted to adapt this program for use in secondary schools by working with secondary teachers.

We found that the opportunity to work with teachers to modify our program was an interesting and valuable experience. In retrospect, we would have done some things differently. In particular, we now feel that it would have more appropriate to have spent considerably more time on procedural aspects of the study than we did (e.g., what the observations were for, how they would be used, when results would be provided to teachers, etc.). More time spent on procedural and social interactions before initiating a focused, decision-making discussion with teachers would have been advantageous. Some initial formal contact with teachers should take place before any substantive discussion, and such meetings should probably give teachers more information about the research and lessen some of their personal concerns about observation and involvement. Such procedural orientation should take place before teachers are asked to read the manual; later, when they have read the manual, they might critique it more fully on substantive grounds.

Another procedure which we would modify involves the initial contact with teachers. Because of the geographical distance involved in this study, this initial contact was made by school administrators. We made some assumptions about the amount of procedural information teachers possessed; however, we found that some teachers did not receive all the information that we thought had been communicated. Also, when working with volunteer teachers, it would be very helpful for investigators to contact teachers and allow teachers to agree (or disagree) about participation, and to negotiate general collaborative arrangements. Some teachers appeared to have volunteered, but without an affective commitment to participation. Investigators could meet with a large group of potential teachers, engage in social and procedural interactions, and only then allow teachers to make a decision about their participation.

In retrospect, the present treatment/discussion meeting may have been too demanding. That is, the teachers were requested to comment upon *all* aspects of the program at a single meeting. If we were repeating the study, we would instead hold two or three different meetings. At the first meeting only development and work on modifications of that aspect of the program would be discussed. In the second meeting seatwork, review, and homework program components would be considered. Such arrangements might lead to a more focused and more thorough evaluation of the program. If problems developed, it might be useful to have yet a third, follow-up meeting to resolve some issues. For example, in the present study, teachers in general had strong feelings that the testing procedures should be systematized, but they had widely varying ideas about how to do this. In retrospect, it seems appropriate to form teacher committees which attempt to develop tentative solutions. Such committees could bring their work back to the whole group for discussion, review, and modification.

Many other changes are also possible in the partnership arrangements; however, the wisdom and desirability of additional strategies depend upon the particular problem being investigated and the types of generalizations which teachers and researchers are trying to make. If investigators are trying to get maximum teacher involvement, it probably is important to bring in videotapes of teaching (particularly tapes which focus on the development portion of the lesson) and to allow teachers to jointly critique and review program components. Teachers could also observe and be observed by fellow teachers in the partnership group so that they could develop fuller understandings of the program and improved strategies for implementing it.

As the data indicate, teacher reactions are likely to vary from site to site (for a variety of reasons), and one critical factor that would have to be considered in any partnership work is how to balance the many demands teachers already face with new demands imposed by project participation. Some teachers would appreciate increased opportunities to interact with other teachers and to discuss the program; however, other teachers may react negatively to extended or involved arrangements. Indeed, it is important for these procedural issues to be resolved by researchers and teachers jointly at their initial meeting so that a common set of expectations about time required for participation (as well as the form of such participation) could be developed.

Because of the context in which we worked (especially time constraints) and the limited amount of time we spent meeting with teachers, we are reasonably pleased with the level of teacher involvement obtained and the ideas which were incorporated into the program. Several program modifications were made, and we think that these changes were appropriate and important for adapting the program to secondary schools. These ideas were essentially teacher initiated, and we are grateful to the participating teachers for their input and assistance. Teachers' brief involvement in training did appear to alter certain aspects of some teachers' behavior, and increased their involvement in the project. However, project involvement did not have a positive effect on other teachers. These individual variations among teachers are similar to results reported by Ebmeier (1978) in the elementary school project. Some teachers in that study implemented the program more fully than others. In particular, Ebmeier found that program implementation was higher among teachers who felt that they were already teaching in ways recommended by the program, and by teachers who were searching for new alternatives. The brief partnership and the general training procedures which we utilized in this study would appear satisfactory for obtaining program implementation from such teachers. However, for teachers who are not interested in seeking alternative solutions and who feel that the program is contrary to their teaching styles, more elaborate procedures and more time will probably be necessary.

7

Teaching and Learning Styles: A Review of the Literature

In previous chapters we have discussed our attempts to find and describe teachers who are able to obtain relatively high levels of mathematics achievement from the class as a class. Now we want to look at teaching and learning from a viewpoint that differs from our previous examination (how general teacher behavior affects general student achievement). More specifically, in the next two chapters we want to discuss different ways of characterizing teachers and students and to consider how different combinations of teachers and students may influence mathematics achievement.

Our purpose in this chapter is to review how others have attempted to define and conduct research on teaching and learning styles. The first portion of this chapter briefly reviews the literature and provides a conceptual framework through which the diverse studies can be organized. Following this initial categorization effort, clusters of individual studies are discussed to illustrate the flavor and complexity of the research field. The last portion of this chapter discusses our own teaching and learning style research within the Missouri Mathematics Program. Here we take an intensive look at our work at the fourth-grade level as a way of introducing the more complex and general findings across three grade levels which are presented in the next chapter.

All of us, both as teachers and as parents, have known children who can listen to or read our instructions and proceed with the task at hand. We have also noticed those who need to be shown how to do a task or watch us demonstrate the process. Why do these differences among children exist and how do they affect instruction and learning? Do certain types of learning modalities jointly affect learning outcomes? Can educational programs be designed to capitalize on students' learning strengths and avoid or remediate weak learning modalities? Unfortunately, we know very little about the various ways in which people intake, process, organize, and express ideas and concepts because most extant research views students as a group with similar response patterns and all amenable to the same instructional techniques.

In an analogous way, teachers commonly have been considered by researchers and administrators as a unit, although common sense would dictate otherwise. Everyone can remember certain professors or teachers whose lectures were spellbinding. Do you remember their counterparts, in whose classes you struggled to keep your eyes open? Some teachers required extensive memorization and recitation of literary works, while others teaching the same course gave open-book tests. Apparently, the same variation found in learning styles is also present in teaching styles.

Unfortunately, our knowledge about the different ways teachers choose to teach, the rationales behind their choices, the variables that affect their instructional patterns, and the interaction of those teaching styles with various students' learning styles is relatively limited. This lack of information is due, in part, to the nature of past educational research, which had tried to isolate singular variables that are useful predictors of effective teaching. By improving a teacher's skills on these selected variables, it was assumed one could improve the general educational program for all students in a class. In the early 1970s when we were beginning our mathematics research, reviews of research were suggesting that this may have been too simplistic a view, since there were few consistent findings (Rosenshine, 1971; Heath and Nielson, 1974) and they yielded little information that could be used to help teachers decide the proper instructional techniques to use for all students across all grade levels and subject areas (Dunkin and Biddle, 1974).

As we have indicated elsewhere in the book, other researchers also became more serious about context as well as the need for systematic research, and their recent research provides some consensus about teaching behaviors related to student achievement in some subjects in elementary school (see for example Brophy, 1979).

Interestingly, because we were aware of these inconsistent and sometimes contradictory earlier findings, we were careful to limit the scope of the Missouri Mathematics Program to a defined context (i.e., fourth-grade mathematics) until we were confident that we had at least some understanding of the effects of our program at that level before exploring other grades. In retrospect, this seems like a wise decision, since we feel that one of the important reasons the Missouri Mathematics Program was able to influence classroom learning positively was the fact that the program was built to suit a particular context. Unlike much earlier research that had attempted to find generalized solutions for teaching across all subject areas and across multiple levels, our program focused specifically on mathematics.

In addition to the realization of the importance of educational contexts, other investigators have suggested that many of the contradictions between research findings and the realities of classroom life may also be the result of ignoring the existence of interactions among teacher characteristics, student characteristics, and instructional strategies. Although the roles of the teacher and student and the purpose of teaching and learning are generally understood, this does not mean that all teachers fulfill their roles in exactly the same manner using the same instructional pattern or that all students approach school with the same desires, skills, and expectations. Each teacher has a unique style based on his or her own knowledge, beliefs, and characteristic patterns of behavior, and each student reacts to the teacher's behavior according to his or her own perception. To understand the behavior of specific students and the behavior of teachers who use certain teaching strategies, it is not enough simply to know the definitions of these roles and the associated expectations; one must also understand the characteristics of individual students and teachers.

An increasing number of researchers (Bush, 1954; Heil, Powel, and Feifer, 1960; Thelen, 1968; Biber, 1958; Newsom, Eischens, and Looft, 1972; Cunningham, 1975; Bennett, 1976; Solomon and Kendall, 1976; Ebmeier and Good, 1979) have suggested that neither the same instructional strategy nor the same teacher type will prove to be optimal for all types of students; rather, one needs to examine the interplay between teacher types and instructional strategies, and their joint effects on various types of students. The term *aptitude-treatment interaction* (ATI) has been used by Cronbach and others (for reviews see Bracht, 1970; Berliner and Cahen, 1973; and Cronbach and Snow, 1977) to describe the systematic attempt to relate individual differences in student aptitude, including aspects of cognitive and affective styles, to instructional methods. If some teachers and/or methods are significantly more effective with certain students than with others, and if all students seem to achieve significantly greater success with at least one type of teacher and/or method, then it may not be possible to specify generic qualities of good teaching. Only qualities that are very broad and self-evident, such as those reported by Rosenshine and Furst (1973) (clarity, flexibility, enthusiasm, task orientation, etc.), or the general concepts developed in the Missouri Mathematics Program such as active teaching and development, would be generally useful for practitioners across many settings.

Because it is difficult to consider simultaneously all of the important factors in interaction research, it is useful to have a model that illustrates the interrelationships among variables. Although the model in Figure 7.1 is not unique and certainly not new, it is nevertheless a useful characterization of the interaction processes among teaching styles, learning styles, and contextual factors. Some portions of the model have been more closely examined than others. For instance, Rosenthal and Jacobson (1968), Brophy and Good (1974), Good (1981), and Cooper (1979) have extensively examined how teachers' past and formative experiences affect their ex-

FIGURE 7.1 A Conceptual Model for Examining the Classroom Interaction Process

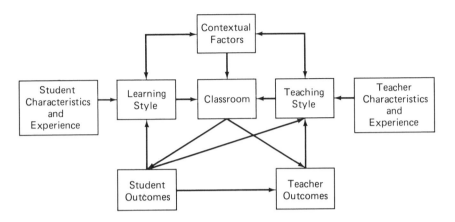

pectations for pupil behavior and ultimately teachers' interactions with students from various ability levels (student types). Similarly, Walberg (1979) and Weinstein (1979) have tried to isolate the effects of various contextual variables on classroom processes. However, our knowledge of most of the interactive processes is still quite limited.

In this chapter we will explore more closely this potentially fruitful but complex area of investigation. Specifically, we will examine how student and teacher types have been formed or determined and on what basis, what similarities these various typologies have in common, and the results of selected studies which have simultaneously examined the interaction of two or more factors of the model. We will also discuss some of our research and findings in these areas.

Student Learning Styles

Although there is general agreement among researchers and practitioners that individual students employ different kinds of learning styles that may vary with task structure,[1] there is no consensus about the definition of the term *learning styles*. As a result, when the term is used in the literature it is conceptually defined in many ways: conditions, content, modes, and expectations (Canfield and Lafferty, 1970); stimuli and contextual elements (Dunn and Dunn, 1978); or many other personality and environmental variables. (See Dunn et al., 1981; and Keefe, 1979, for more comprehensive reviews.) The term learning style, then, is a rubric for a

---

1. Task structure refers to the sequencing and difficulty of the subcomponents of the overall task (i.e., multiplication would be a subcomponent or necessary step in division).

host of conceptions about the different ways in which children receive, process, and respond to various kinds of information.

Considering the multitude of possible personality, contextual, and process variables which are involved in learning, the task of narrowing the field to a manageable number that account for a significant portion of the variation among students is truly difficult.

Fortunately, diverse student traits upon which learning styles are ultimately based can be generally classified by the three-dimensional system in Figure 7.2, which examines *trait type, trait dimension*, and *trait domain*. Trait type refers to the degree to which the identified variables are genetically/environmentally dependent. In the most restrictive sense, these would include student types formed on the basis of sex, age, or some physical characteristics. Closely related would be neurological brain behavior (Thies, 1979), followed by various ways that the mind processes information, such as cognitive style, conceptual level, etc. (Coop and Sigel, 1971; Hunt, 1971; Kolb, 1981; Gregorie, 1979; Hill, 1971). At the other end of the continuum are student variables that are largely socially determined, such as emotional maturation, persistence, responsibility, and need for structure (Dunn and Dunn, 1978).

Trait dimension refers to the scope of a trait. A trait can be singular (e.g., field dependency), or multi-, or megadimensional (e.g., motivation

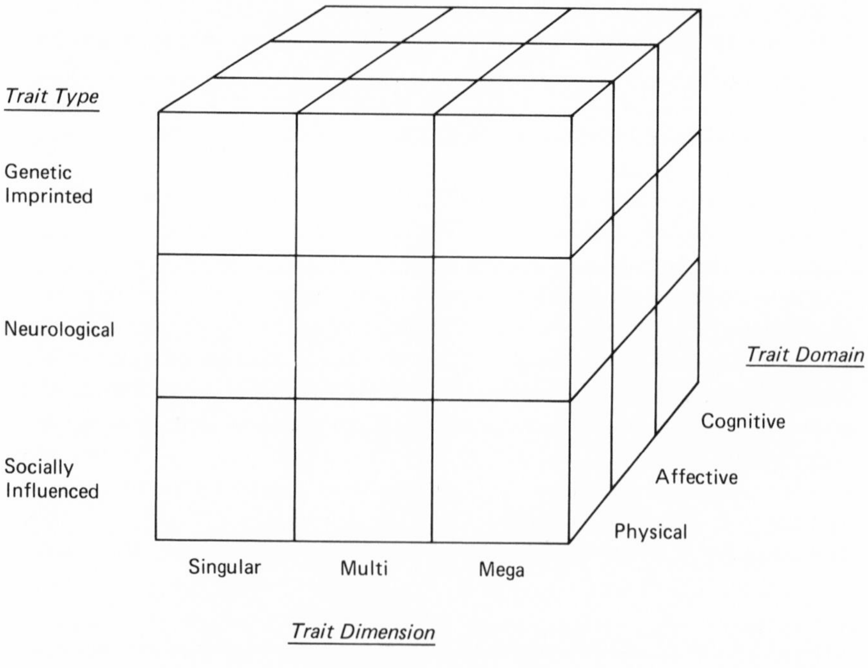

FIGURE 7.2   A Conceptual Classification System for Teacher and Student Traits

and persistence). Most ATI work has focused on singular dimensions for three reasons. First, it is conceptually easier to interpret results and establish causation when the number of variables is relatively small. Every dimension added to an experimental design increases the number of possible interaction patterns and often makes the accumulation of adequate sample sizes and the meaningful interpretation of results impossible tasks. Second, many ATI researchers have followed tradition in educational research by searching for a singular variable that has a profound influence. Although they have broadened their perspective by including various levels of the treatment, the prodigious move to a multidimensional causative model has not yet been made. Finally, although there is clearly a need to understand student learning styles from a broad perspective, many researchers do not have the technical knowledge necessary to analyze relatively complex designs. Though it certainly would not be advisable to abandon all research that investigates unidimensional traits (for much work needs to be done here), researchers probably at least need to consider the overall context of schooling and deal with variables or clusters of variables within the instructional setting that have meaning to teachers.

A third dimension of student and teacher traits is trait domain, which can be divided into three areas: cognitive, affective, and physiological. (See Keefe (1979) for a more complete discussion.) The cognitive area includes academic processing modes or potential ways of perceiving, organizing, analyzing, evaluating, synthesizing, processing, and thinking about information. As could be easily deduced, the majority of ATI research has concentrated on this area. Although some authors choose to separate cognitive styles from abilities, there is no intent here to make such a distinction, because the two dimensions overlap. Rather, it seems more useful to include aptitude dimensions (IQ, mental ability, scholastic ability, etc.) in the general cognitive category, and to subdivide the entire domain into three dimensions: aptitude, reception, and retention. Some exemplars of the reception dimension are sensory, kinesthetic or psychomotor, visual, and auditory modes (Sperry, 1972); differences in the way individuals deploy attention (Holzman and Klein, 1954); and susceptibility to distraction and distortion of tasks (Gardner et al., 1959). The concept formation and retention dimension could be represented by the cognitive complexity work of Harvey (1967), which focuses on understanding the differences in the number of dimensions utilized by individuals to construe the world.

The second trait domain encompasses dimensions of personality that are related to affective behavior. This domain can logically be subdivided into three types of affective reactions: toward the environment, toward self, and toward others. Conceptual levels (Hunt, 1971) are good examples of the affective reaction to the environment trait characterizing how much structure a student requires to facilitate optimal learning. Hunt's definition is based on a developmental personality theory that describes persons in terms of a hierarchy of increasing conceptual complexity, self-responsibility, and independence. A low conceptual level indicates a need for high structure, while the reverse is true for persons identified as possessing a high conceptual level. Dunn and Dunn (1978) also include environmental elements in their definition of learning styles (need for appropriate levels of sound, light, temperature, and structural design, etc.).

The second subcategory of affective-domain traits can be described as the learner's or teacher's perception of and attitude toward himself and would include a multitude of variables from psychology, sociology, and psychiatry. Any comprehensive discussion of self-concept would result in several volumes and is beyond the scope of this book. However, it is useful to mention a few areas that have received persistent attention, relate directly to classroom processes, and seemingly offer the most potential for further research. One topic that has been extensively researched is locus of control, which includes students', and sometimes teachers', perceptions of causality in behavioral outcomes on a continuum from internal to external. Internal persons think of themselves as responsible for their own behavior, as deserving of praise for successes and blame for failures. External persons tend to see what happens to them as the result of circumstances beyond their control, as due to chance or luck. Achievement motivation (McClellan, 1972; Ebmeier, 1978) and persistence (Carroll, 1963; Dunn and Dunn, 1978) could also be classified as affective traits and are thought by most educators to be important student variables related to the instructional process. Teachers who know how to deal with students low in persistence and/or motivation in a regular classroom environment and who understand the causes of such student behaviors can improve instruction significantly. Other affective traits might include risk taking (Kagan, 1972), level of aspiration (Bloom, 1976), and social motivation (Hill, 1971; Bennett, 1976).

One interesting aspect of self-concept and other affective traits is that they can be classified as dependent variables as easily as categorical variables. The independent effect of a treatment on the variable initially used for categorization purposes might be of more interest than the treatment effect per se. For instance, a treatment program designed to improve mathematical skills of students with low achievement motivation might also increase the motivation of the participants, in addition to improving mathematic performance. In this instance, the improved motivation would be an important outcome as well.

The last logical subdivision of affective traits concerns students' and teachers' reactions to others. Since many learning activities, particularly in the lower grades, are dependent upon cooperation and sharing, students' perceptions of and reactions to peers and adults may be of significant importance to their learning. Typical of these traits are Canfield and

Lafferty's (1970) "academic conditions," which include "relations with instructor and peers"; Dunn and Dunn's (1978) sociological "dimension," which rates students in terms of their reactions to various grouping patterns; and Ebmeier's (1978) "other orientation," which describes individual students' general reactions to peers, teachers, and significant others. This category is not only conceptually practical, but its applications in terms of grouping patterns and task assignments are widespread, because most teachers have a significant amount of control over seating patterns and work arrangements.

In terms of formation of teacher and student typologies, the affective domain has been researched and written about more than the physical or cognitive domains although the affective domain is a very complex area due mainly to the instability of many human affective responses. For investigators, the attractiveness of this area stems in part from the perceived ease of data collection, usually via attitude surveys. However, some authors have repeatedly pointed out that reliance on students to identify their own learning styles is certainly questionable (Fischer and Fischer, 1979). Even though self-report attitude surveys may serve a useful purpose in exploratory studies, until actual observational data or other confirmatory procedures can be undertaken, student questionnaire data should be only tentatively accepted.

A final domain, physical, that can be useful in describing the vast number of possible traits for potential investigation use in ATI studies is derived from consideration of the normal functioning of the human body. Variables such as gender, left or right hand dominance, etc., are among the most evident physical influences on school learning. Recognition and accommodation of the physical differences among students are most prominent in special education classes in which sophisticated equipment has been developed to provide for various student learning styles which are dictated by physical characteristics. Although the vast body of special education literature relating to individual differences will not be reviewed here, it is important to note that physical factors have long been recognized in this field as an important factor in designing educational programs.

Selection of Traits for Study. From the previous review of trait types, dimensions, and domains, it is clear that the choice of appropriate variables for ATI studies is a formidable task, for every variable cannot be considered. As one would expect, the selection criteria used to identify the most useful variables for study have been as diverse as the number of possible traits which can be studied. However, most variables are selected for study on the basis of one or more of four overlapping criteria. The first and most common criterion is simple availability. Data from school records, natural classroom grouping patterns, test results, former teacher recommendations, and easily observed physical characteristics of students have served for years as means for discriminating among students. Though this method may lack some rigor, it is the most common and is utilized by all teachers when they group for instructional purposes.

The isolation of traits based on a particular conceptual theory is the second most common method of variable selection (Hunt, 1977; Kolb, 1981; Schmeck et al., 1977; etc.), although the definition of theory may vary considerably. In some cases the theory is based on sound empirical investigations and carefully conceived models, while in other instances common sense and personal opinions seem to serve as the base for the theory.

The third selection criterion results from process/product investigations with their strong emphasis on naturalistic classroom observation. The variables chosen usually evolve from classroom observation and may vary across contexts. (See Brophy and Evertson, 1981, for a further discussion.) The final selection method employs some form of multivariant analysis to mathematically derive salient factors from a host of possible variables (Bennett, 1976; Solomon and Kendall, 1976; Cunningham, 1976).

While the use of various selection criteria and procedures may help reduce the number and complexity of variables under consideration, the number of resultant variables may still be unacceptably high for meaningful analysis. To solve this dilemma of either loss of information by further reduction in the number of variables or the inclusion of so many independent measures or factors as to render interpretation impossible, most researchers have used either some form of profile analysis and/or formed a finite number of student typologies. In either case, an attempt is made to isolate consistent patterns of traits and to examine how these patterns interact with treatments or contexts. The rationale for employing such group procedures is that they avoid the artificiality of treating each trait as an isolated variable (ignoring possible joint or confounding effects of other variables), or of attempting to account for the simultaneous effects of numerous variables with complex, often uninterpretable interactions or with elaborate attempts to control statistically factors which are not independent in nature.

Student Type Formation. For the sake of conceptual clarity, most authors have discussed learning styles in terms of specific characteristics or clusters of characteristics, which in a broad sense could be termed student typologies. In the simplest case these typologies are based on measurement of a single trait. All other factors are ignored or held constant either experimentally or statistically. Examples of this approach would be the work of Peterson (1977, 1981), who classified students in terms of need of structure and class participation.

There are seemingly three clear advantages to limiting the variables which constitute typologies: (1) characteristics of the typologies are clear and understandable; (2) treatments are relatively easily designed to match these defined characteristics; and (3) results can be attributed to a small number of learner traits interacting with one treatment. Although studies employing a small number of independent variables are important for development of theoretical constructs, such findings do not have clear implications for teachers, who must work in a dynamic environment with classes composed of children who have many different traits which interact.

Four methods have generally been employed by researchers to derive multiple-trait measures, depending on the nature and magnitude of the traits involved. The derivation of student typologies from students' *cognitive styles*, although not truly based on multiple measures, is one method of typology formation which utilizes a broad definition of the trait concept (e.g., as opposed to IQ or sex). The term cognitive style has been used most frequently to denote consistencies in individual modes of functioning in a variety of behavioral situations. Witkin (1962) has defined cognitive styles in terms of field independence and field dependence. Broverman (1980) spoke of conceptual vs. perceptual dominance. Gardner (1959) studied cognitive control patterns that suggest a type of cognitive and/or perceptual style, such as leveling and sharpening, field articulation, and equivalence control. Kagan, Moss, and Sigel (1963) described three basic cognitive styles—descriptive, relational, and categorical—based on children's performances on grouping and sorting tasks. Finally, some writers have even applied the label of cognitive style to Guilford's (1959) convergent, divergent, and evaluative types of cognitive operations.

The use of the term cognitive style to denote diverse referents has led to some confusion on the part of those who read the research literature. The lack of agreement on a single definition has caused this term to become investigator-specific, and therefore difficult to use and report readily in ATI research. The advantage in using cognitive style, however, is that it generally includes a broader sampling of student traits. While this makes it more difficult to conceptually define the typologies, the results of the research are more applicable to school settings.

The second method of deriving types of students from multiple traits is based on *learning style inventories*, chiefly those of Canfield and Lafferty (1970); Dunn, Dunn, and Price (1975); Gregorie (1979); Kolb (1981); Moos (1978); and Schmeck (1977). Although in most cases student typologies are not inherent in the inventories, nor do the authors suggest appropriate ways to cluster students based on their instruments, the resultant student profiles can easily be used to group students according to the similarity of their responses. The derivation of student profiles from a wide variety of variables represents an important step for ATI researchers because they are now likely to get a more complete representation of the "total student" than they would by using individual characteristics or cognitive styles to form student typologies. However, there are two disadvantages of using learning styles to categorize students. First, researchers can never be completely certain whether (and to what extent) the traits selected actually represent relevant dimensions of student behavior. Second, as is true with the use of cognitive styles, the further one deviates from singular measures, the more abstract the dimensions of the derived typologies become.

The third method of typology formation, *mathematical derivation*, has been largely dependent on the development of computer programs which can manipulate large amounts of data. Since these programs have only recently become widely available, few studies have employed this method. The procedure usually involves the administration of a large variety of student measures, followed by some form of multivariant data reduction technique which mathematically derives the typologies based on similar responses to the traits chosen (Power, 1974; Britt, 1971; Solomon and Kendall, 1976; Bennett, 1976; and others). As with learning style inventories, the chief problem associated with mathematical derivation is that as one moves away from singular trait dimensions the task of defining typologies becomes more difficult. The advantage, of course, is that the typologies are based on more comprehensive data.

A fourth method of student type formation is more subjective than those previously discussed in that it *relies on teacher/observer opinions* about what student traits belong together and thus define a typology. For example, Jackson et al. (1969) utilized teachers' descriptions to identify four student clusters: attached, indifferent, concerned, and rejected. Attached students were described as conforming to institutional and teacher expectations, causing few behavioral problems, and making few demands on the instructor. Concerned students made extensive but appropriate demands upon the teacher, were very cooperative, and worked willingly. Indifferent students were seldom involved with the teacher and were normally ignored, while the rejected students were perceived by teachers as being overwhelming or impossible.

Good and Power (1976) described five types of students which emerged from both empirical and subjective analysis: success students, social students, dependent students, alienated students, and phantom students. Three of Good and Power's derived types closely correspond to three of the student types reported by Jackson. Jackson's description of the rejected student is similar to Good and Power's alienated student in that these student types are seen as overwhelming and frequent causes of behavioral problems within the class. Good and Power's dependent student and Jackson's concerned student also share many characteristics typical of anxious/acceptance individuals. The attached and success student types are both high achievers, active, and independent.

Because these student typologies are formed at least in part from subjective personal opinion, the final placement of students into one of the derived typologies must also be done in a subjective way, usually through teacher identification. Beckerman (1981), for example, asked teachers to identify which of their students clearly fell into one of the student categories previously defined by Good and Power (1976). This procedure of student placement has distinct advantages in that it avoids some of the scaling problems inherent when similar instruments are used across grade levels to type students, and hopefully incorporates information based on classroom observation through the teacher's choices. However, the reliability and validity of the teachers' categorizations remain to be demonstrated. As an example of the problems of nomination typology formation, Beckerman was able to use less than half of the available cases in his study because of teachers' reluctance to classify students and the unreliability of teacher classifications compared to other information. Such problems severely hamper the generalizability of results.

Teaching Styles

The ways in which instructors teach their classes can have important consequences for students' learning, satisfaction, and development. Teaching styles reflect teachers' educational values as well as the goals they hope students will attain. This fact is not very revealing, because it has been long recognized that variations exist in the way teachers approach their tasks (as evidenced by the hundreds of studies which have attempted to link teacher characteristics and instructional patterns to student outcomes). Although there is a massive body of research concerning the ways teachers gather materials, construct teaching plans, and present lessons, the concept of teaching styles has received relatively little direct research attention. Just as there are many learning styles, there seem to be a variety of identifiable styles of teaching. (See Brophy, 1982; Shavelson and Stern, 1981.)

As with student learning styles, teacher typologies have been derived in a number of ways using varied traits, and teaching styles can be classified in a manner similar to student learning styles. Researchers have attempted to describe teaching styles by such varied methods as systematic observation (see the review by Medley and Mitzel, 1963), rating methods (Remmers, 1963), and measures of social interaction (Withall and Lewis, 1963). More recently, the classroom environment has been studied by asking students about their perceptions of classroom procedures and behaviors (e.g., the May, 1982, issue of the *Elementary School Journal*).

The most direct method of forming teacher typologies involves selecting a singular trait (i.e., sex or intelligence) and forming groups based on the presence of this trait. Since the research literature is abundant in these areas and reviews are available, individual studies will not be discussed here; rather, we will focus on illustrative studies which have employed multiple-trait measures. In terms of the classification systems in Figure 7.2, almost all existing studies of teaching style employ a socially determined affective-trait type (as opposed to a mental process such as IQ), and most are megadimensional. No studies were located in which the selected teacher traits used for forming typologies were related to genetic characteristics from the cognitive or physical trait domains.

Baird (1973) used a 33-item student questionnaire dealing with teaching practices such as examination, classroom procedures, student-teacher interaction, assignments, and instructor attitudes, to explore the utility of a model consisting of six dimensions of teaching style: the didactic, which emphasized specific, detailed knowledge of facts and comprehensive coverage; the generalist, which was concerned with helping students apply ideas and facts; the researcher, which placed importance on the interpretation and analysis of information; a dimension which concentrated on the relevancy of student responses to classroom activity; a dimension which concerned the relative clarity or ambiguity of the teacher's expectations; and a sixth type which emphasized the affective rewards given by a teacher.

In a study employing a trait from the cognitive domain, Murphy and Brown (1970) used the Harvey, Hunt, and Schroder (1961) conceptual system, which postulated four stages of conceptual development ranging from concreteness to abstractness, to place teachers into typologies. They administered Harvey's Conceptual Systems Test (1967) that measured six factors (divine fate control, need for structure-order, need to help people, need for people, interpersonal aggression, and anomie) to 136 student teachers and formed four teacher typologies based on the results. Type 1 teachers were characterized by dependence on authority sources, and they regarded what was stated by these authorities as the ultimate word. Type 2 teachers tended to reject the customary social roots for self-definition and esteem, and lacked stable referents for their concepts, which created inconsistencies and uncertainties in their functioning. Teachers characterized by conditional dependence and having high affiliative needs were grouped into Type 3, while Type 4 teachers were independent, having conceptual structures which were cognitively complex.

In studies where the nomination method of teacher trait selection is used, investigators simply propose typologies based on their subjective judgment of important patterns of traits. For example, Fischer and Fischer (1979) identified six teaching styles (task oriented, cooperative planner, child centered, subject centered, learning centered, and emotionally exciting and its counterpart) based on experienced observation and reflection.

Investigators (Solomon, 1976, and Bennett, 1976) utilizing data reduction techniques (cluster analysis, discriminant analysis, etc.) generally collect a wide variety of teacher traits, and then isolate dimensions of teacher behavior after examining the data. Once these dimensions are established and defined, a teacher's composite scores on each dimension are used to develop an individual profile and ultimately to cluster teachers with similar profiles into groups.

Both nomination and data reduction methodologies have advantages and disadvantages. The former, while viewing teaching from a more molar framework, are dependent upon the intuition of the typology developer and, hence, may be merely a reflection of the researcher's biases. The relative ease of composing student and teacher typologies through the nomination method has resulted in identification of teaching and learning styles which have little conceptual or empirical basis. The data reduction method also presents one serious problem in that it is often difficult to conceptually define the nature of the derived dimensions of teacher behavior. If the dimensions are poorly characterized, then defining the teacher or student typologies which are based on the derived dimensions is difficult.

Illustrative Interaction Studies

It seems probable that there would be a significant number of empirical studies which have examined these interactions. However, beyond those studies which have investigated singular traits, there are only a few interaction studies: Crist and Hawley (1976); Bennett (1976); Solomon and Kendall (1976); Cunningham (1975); Moos (1978); Ebmeier and Good (1979); and Beckerman (1981).

Crist and Hawley investigated the relationship between cognitive and affective characteristics and mathematics achievement for sixth graders, using three instructional treatments in one school district. Self-concept, locus of control, field dependence-independence, ability, and previous achievement level represented learner characteristics. The three mathematics treatments were: traditional whole-class instruction based on a basal textbook; a locally developed mathematics management system which correlated instructional objectives with various published curricula utilizing a combination of individual, small-group, and large-group instruction; and individually prescribed instruction. The presence of each treatment was verified by classroom observations. Math achievement was measured by the three subtests of the Comprehensive Test of Basic Skills (Computation, Concepts, and Applications), and the results were submitted to regression analysis.

The results of this study showed that students have varying patterns of ability, achievement, and personality traits which interact in complex ways. For the particular sample in this study, pretest achievement appeared to be the most salient aptitude characteristic which interacted with the treatments to influence achievement outcomes (although field dependence-independence also interacted to a large degree). There was evidence that students at higher pretest achievement levels performed better on the Computation Subtest and the composite Arithmetic Total in traditional instruction than in the individually prescribed instruction treatment, and that students with lower pretest achievement scores achieved better with individually prescribed instruction. However, overall, students who had highest the math achievement had received whole-class instruction and the lowest scores were generally obtained by students who had individually prescribed instruction.

Cunningham (1975) examined the student type–teacher type interaction between 108 kindergarten teachers and their 1,291 students. He hypothesized that various types of students tend to receive different benefits from various types of teachers. Teacher characteristics included age, sex, race, years of teaching experience, and a teacher belief survey which was composed of six subscales that measured six distinct dimensions of teacher beliefs. A Q-mode factor analysis was used to identify types of teachers. Four factors were found to account for 94.5 percent of the variability in the measures used to characterize the 108 teachers:

1. White/subject integration
2. Black/experienced
3. Inexperienced/student-centered/empathic
4. White/experienced/subject-centered

Teachers were classified as members of a type only when their characteristics were similar within type and different from the other categories. Participants who could not be classified clearly were dropped from the study. Unfortunately, Cunningham did not indicate how many were dropped or verify the existence of these typologies in the classroom.

Cunningham's criterion variables were selected to match the objectives of the kindergarten program. These were defined as the differences between observed and predicted scores on specific language, mathematics, and task-orientation tests. The measures used to describe the students were age, sex, race, a home information scale, a measure of preschool achievement, intellectual level, language and math concept development, introversion-extroversion, negative-positive social behavior, and positive-negative task orientation. Four factors accounted for 94.4 percent of the variability in the measures used:

1. Young/advantaged
2. Extroverted/black/female
3. Introverted/disadvantaged/white/female
4. Slow/alienated/male

Of the 1,291 students, 720 were classified as clearly falling into one of the four types; the rest were dropped from the study. Whether the high number of students dropped from the study resulted from inaccurate student typologies or from the use of very strict guidelines when placing students into one of the four student types is unclear.

The results of the study indicated that the four teacher types identified were not equally effective with all types of students. Cunningham reported that certain types of teachers were significantly more effective with one type of student than with another and that all student types seemed to have at least one type of teacher with whom they could achieve significantly higher. In addition, examination of student and teacher profiles revealed that students whose descriptive profiles differed from one another most tended to vary most in the benefit they received from a particular teacher type. This finding implies that there might be some continuous mathematical relationship between student type, teacher type, and achievement, so that the closer the match between the ideal student type and associated teacher type the higher achievement would be.

Solomon and Kendall's (1976) research parallels that of Cunningham, with one exception. They view teacher types as a subset of a larger dimension they refer to as classroom type. Their research is therefore reported as classroom types, although classroom types and teacher types may be thought of as overlapping concepts. The selection of one term or the other depends on the variables used to define the typologies and the researcher's preference. Solomon and Kendall's sample consisted of 50 individual fourth-grade classrooms. To gather information about classrooms, they employed a systematic observation system and a teacher questionnaire. The observation instrument included measurements of general classroom activities, classroom atmosphere, teacher activities, and student activities. Trained observers visited each of the 50 classrooms in the study for one-hour observation periods. The visits were distributed throughout the school year and were balanced between morning and afternoon and among different days of the week. In order to obtain teachers' views of the characteristics, organization, and typical activities of their classrooms, they were asked to respond to a 64-item questionnaire.

Separate factor analyses were conducted with each section of the observation form and with the teacher questionnaire. These analyses produced a total of 33 factors. Factor scores from this initial procedure were used as input in a second-order factor analysis which ultimately produced six factors considered to represent basic dimensions of classroom organization and activity. The 50 classrooms were then cluster analyzed into six groups based on the similarity of their profiles on the six factors. Types were described in terms of the average profile of all the classes which constituted that cluster. Two examples follow.

Cluster 1 classrooms were extremely permissive; lacked control and orderliness; had varied, student-initiated activities; were moderately warm; and tended to have individualized interaction between teachers and students. Although they showed some of the characteristics which have been attributed to open classrooms, their extreme lack of control and order was beyond that recommended for the ideal open classroom.

Cluster 2 classrooms were highly controlled and orderly, but students also had relatively frequent opportunities to initiate their own, varied activities. These classes had little interaction between teachers and individual students and tended to be relatively cold. Students usually directed their own activities, but in a structured and somewhat impersonal setting.

Student types were formed in a manner analogous to the formation of classroom types. Tests measuring creativity, inquiry skill, self-esteem, attitudinal orientation toward school, achievement, motives, preferences, and orientations were administered at the beginning of the year. Each of these sets of tests and teacher ratings concerning the classroom behavior of students in their classes was factor analyzed. Eleven factors or underlying dimensions were produced. Clusters of students were derived based on the similarity of their profiles on these 11 dimensions. This cluster analysis produced three distinct student types, each with a different central profile.

Type 1: These students were described as low achievers who were not intrinsically motivated. They were inner oriented, lacked self-confidence, rated high on compliant-conforming orientation, and were moderately achievement motivated and self-directed.

Type 2: These students tended to be highly motivated, self-confident, prior achievers. They scored low on self-direction and preference for autonomy, and were moderately compliant.

Type 3: These students stated strong preferences for autonomy, personal expression, and self-direction. They scored low on compliant orientation. Their prior achievement and motivation scores were moderate, except for achievement motivation, which was low.

To assess the student type–classroom type interaction, Solomon and Kendall employed 14 dependent measures. These included achievement (reading, mathematics, and language), creativity, inquiry skill, writing quality, four attitude and value factors, and a measure of self-esteem. Two factors (perseverance/social maturity and activity/curiosity) were derived from teachers' ratings, and the remaining three factors (enjoyment of class, social involvement, and perceived disruptiveness in class) resulted from the students' self- and class evaluations.

Solomon and Kendall employed a repeated-measures analysis of variance procedure as a framework for data analysis. Each classroom was treated as if it were a single subject. The interesting subcategories within a classroom were treated as representing repeated measurements of the same subject, made under different conditions. Significant two-way interactions between student types and classroom types were found when creativity and activity/curiosity were used as dependent variables. When sex was partitioned out as a third variable interacting with student type and classroom type, significant three-way interactions were found on self-esteem and four attitude and value dependent variables. No significant interaction between student type and classroom type was found when using a composite measure of achievement (reading, mathematics, and language).

Unfortunately, interactions between student and classroom types on each subscale of the composite achievement dependent variable were not reported. Using composite scores on achievement may mask important student type–classroom type interactions. For instance, two of the subscales of the composite achievement scale (reading and language) are highly dependent on home environment factors; thus, one would not expect as much interaction with classroom variables as with the mathematics subscale. If there were, in reality, significant interactions between student types and classroom types on mathematics achievement, they might never be discovered, since inclusion of the other two dependent factors (reading and language) would mask the interactions. This could possibly explain the lack of interactions between student type and classroom type on the achievement dependent variable in Solomon and Kendall's study.

Bennett (1976) carefully compared the effects of formal, mixed, and informal teaching upon students' performance on a standardized achievement test, ability to write stories, and their affective reaction to schooling. Teacher types and student types were developed, and the interactions between them on the dependent variables were examined. To determine teacher types, classroom teachers from 12 schools were interviewed to identify teaching behaviors that differentiated progressive from traditional teaching style. Eleven differentiating elements were isolated in these interviews and translated into questionnaire items about classroom behaviors in six major areas: teacher control and sanctions, classroom management and organization, curriculum content and planning, instructional strategies, motivational techniques, and assessment procedures. Ultimately, teacher types were formed on the basis of the questionnaire responses of 468 fourth-grade teachers. The clustering procedure yielded 12 teaching styles; two are described below.

Type 1: These teachers favor integration of subject matter and allow pupil choice of work, individually or in groups. Most allow pupils to choose seating. Less than half restrict movement and talk. Assessment in all its forms—tests, grading, and homework—appears to be discouraged. Intrinsic motivation is favored.

Type 2: All members of this group emphasize teaching subjects separately by whole-class teaching and individual work. Pupil choice of work is minimal, although most teachers allow choice in seating. Movement and talk are restricted and offenders are often physically disciplined.

Teacher typologies were validated by classroom observation, reports of primary school advisors, and a content analysis of student reports. This tripartite validation was important to insure that derived typologies were distinct and represented real differences in classroom teaching behavior.

Bennett's student typologies were derived from a proposed model of personality and behavior in school. The model suggests that most behavior can be explained in terms of two dimensions: adjustment-maladjustment and assertion-compliance. A wide variety of personality tests were administered and the results cluster analyzed to yield eight characteristics of pupil personalities. These include extroversion, neuroticism, conscientiousness, self-evaluation, anxiety, motivation, associability, and conformity. Student scores were then cluster analyzed in terms of these eight dimensions, resulting in eight pupil typologies, two of which follow.

Type 2: These students are very introverted and neurotic, with a high degree of classroom anxiety. They hold a poor view of their ability and tend to be both unsociable and nonconforming.

Type 8: These are stable, extroverted, motivated, sociable, conforming students who hold favorable self-images.

To examine the interactive effects of teacher and student types, measures of achievement in reading, mathematics, and English were collected in 13 relatively informal classrooms (Teacher Types 1, 2), 12 mixed classrooms (Teacher Types 3, 4, 7), and 12 formal classrooms (Teacher Types 11, 12). Analyses showed that students of the same personality type progress differentially under different teaching styles. The effects of teaching style, however, were much more powerful than the effects of student personality. Most students achieved better in formal classrooms, especially in mathematics.

The probability of finding interactions between student and teacher types is obviously limited by the nature of the data from which the typologies are formed. Although interactions were found in this study, stronger interaction patterns might have been discovered if Bennett had included aptitude data in forming the student typologies. Teacher typologies might have been strengthened if they had been based in part on data concerning characteristics that differentiate one teacher from another in general rather than only on characteristics that separated traditional from progressive teachers. Additionally, it would have been advantageous to leave the 12 teacher typologies intact for the data analysis rather than collapsing them into three categories (formal, mixed, informal). This collapsing may have masked significant interactions between teacher types and student types.

A CASE STUDY OF TYPOLOGY INTERACTIONS IN THE FOURTH-GRADE MISSOURI MATHEMATICS PROJECT

In the previous section of this chapter we discussed some individual studies that have examined the interactive effects of teaching and learning styles on subsequent achievement. In addition, we presented a conceptual framework around which the divergent studies could be organized and discussed. The purpose of the remaining portion of this chapter is to take an intensive look at our work in this area. However, rather than attempting to simultaneously discuss the multitude of findings across three grade levels, we thought it would be better to select one study from our ATI work at the fourth-grade level to give the reader a better understanding of the general method and findings. The information from other grade levels is presented in the next chapter.

As part of our interest in effective strategies of teaching mathematics we were concerned about the effect of our treatment program on different types of teachers and students. In particular, we were curious about the possible differential effects the program might have on students with varying aptitudes and/or abilities and therefore included a student-type dimension in the fourth-grade experimental study. The choice of aptitude variables on which to gather data was difficult, and we spent much time considering one or more dimensions. Ultimately, we decided to select student aptitude variables that we thought might interact the key features of the treatment program. For instance, because the experimental treatment program called for daily assignment of homework, we included a student-aptitude dimension termed *conscientiousness*, which we believed would be an important predictor of homework completion. In terms of the classification system in Figure 7.2, the student-aptitude variables selected could be described as overt behaviors (trait type) that are multidimensional (trait dimension) and primarily from the affective domain (trait domain). These aptitude traits are described below.

1. *Mental Computations*: Like/dislike of doing mental computations independent of pencil and paper mnemonic devices.
2. *Conscientiousness*: Forms of conscientiousness, such as completion of homework, keeping track of papers, and remembering what to do.

3. *Choice*: Preference for choice in assignments and activities in math class.
4. *Dependence*: Dependence on the teacher for initial structuring of the math lesson.
5. *Other Orientation*: Like/dislike of working with others to solve math problems.
6. *External Motivation*: Dependence on external forces such as checking of papers for motivation in math.
7. *Misbehavior*: Amount of trouble the child gets into in school.

To obtain student aptitude data a self-report instrument was devised and administered in the early fall in the experimental study (See Ebmeier and Good, 1978, for a discussion of the metric properties of the instrument). Students were mathematically clustered into groups according to their responses to these subscales, sex, and preexperiment achievement test results. Profiles of these typologies appear in Figure 7.3 and Table 7.1, while verbal descriptions are presented below. Again, we reiterate that it is difficult to express concisely complex "types" that have been derived mathematically.

Student Type 1—Dependent. Students in this cluster scored slightly below the average on prior math achievement. They displayed an average conscientiousness in completion of papers and assignments and needed some adult encouragement to complete their work. Behavioral problems were reported as moderately low in frequency. In general, Type 1 students could be classified as "typical" in most respects. The key character-

FIGURE 7.3 Graphic Representation of the Standardized Scores for Each of the Four Student Types (Grade 4)

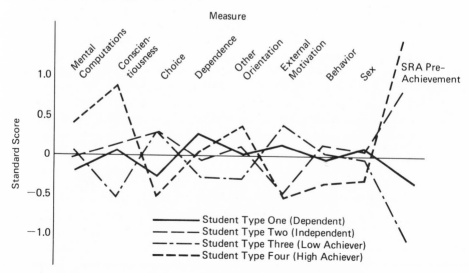

TABLE 7.1 Student Typologies Based on the Cluster Analysis: Means, Standard Deviations, and F Ratios for Cluster Components (Grade 4)

Student Typology		Mental Computations	Conscientiousness	Choice	Dependence	Other Orientation	External Motivation	Behavior	Sex 1 = Male 2 = Female	SRA Pre-achievement Score
One (N = 388)	Mean	2.46	7.32	1.11	4.44	1.87	4.55	0.50	1.55	10.97
	S.D.	1.47	1.87	1.22	0.69	1.18	1.23	0.83	0.49	1.80
Two (N = 214)	Mean	2.69	7.40	1.89	4.09	1.95	3.66	0.70	1.55	17.19
	S.D.	1.48	2.07	1.51	1.18	1.39	1.65	0.97	0.49	3.22
Three (N = 344)	Mean	2.72	6.13	1.90	3.91	1.50	4.90	0.65	1.49	6.70
	S.D.	1.35	2.29	1.47	1.10	1.10	1.15	0.91	0.50	2.24
Four (N = 151)	Mean	3.29	8.82	0.78	4.18	2.24	3.66	0.27	1.35	20.86
	S.D.	1.19	1.08	0.98	0.96	1.32	1.52	0.70	0.48	5.62
Total (N = 1097)	Mean	2.70	7.17	1.46	4.17	1.82	4.36	0.55	1.50	12.21
	S.D.	1.40	1.97	1.33	0.97	1.22	1.34	0.87	0.49	3.01
F Ratios (3,1093 df)		12.60*	68.60*	41.08*	18.71*	14.55*	54.37*	8.95*	6.41*	1009.42*

*p ≤ 0.001.

istic that distinguishes Type 1 students from all other types is that they scored very high by comparison on the dependence scale, which suggests that they are very dependent on the teacher for direction and guidance.

Student Type 2—Independent. Students in this typology scored considerably above average on prior math achievement. They tended to be about average on mental computations, conscientiousness, dependency, and other orientation. Scores on the choice (high) and external motivation (low) scales indicated that students in typology two might be classified as independent. Their scores on the behavior scale indicated that these students are the most frequent behavioral problems of the four student types. This seems to mesh with the independent label, because one could expect students who are slightly unconventional and who like a wide latitude of choice to clash with a teacher's idea of a well-behaved class. The moderate, although below average, score on the dependence scale is somewhat puzzling. It appears that although these students prefer a wide latitude of choice and are not especially affected by a teacher holding them accountable (i.e., by saying, "I'll check this work at the end of the period."), they seemingly still rely on the teacher to provide initial structuring of the lesson.

Student Type 3—Low Achievement/Withdrawn. Clearly the most salient feature of students in Type 3 is their extremely low prior math achievement. Scores on the other scales reflect what one would expect from low achievers; low conscientiousness, low on teacher structuring dependence, low on other orientation, high on need for external encouragement to finish their work, and above average on behavioral problems. Whether those traits cause, occur simultaneously, or are a result of low achievement is unknown. They do, however, appear to cluster together very nicely. Students in Type 3 appear to be somewhat withdrawn from the main flow of classroom life. For instance, they do not like to work with other children, nor do they especially depend on the teacher for structure. Their high preference for choice is possibly an attempt to avoid problems or situations that might be aversive, since they would likely fail most academic tasks that the teacher might provide.

Student Type 4—High Achievement/Task Oriented. Students constituting Type 4 are opposite in almost all traits from Type 3 students. They are very high achievers who are conscientious, other oriented, require little external motivation to complete assignments, like to work problems in their heads, are infrequent behavioral problems, and are more likely to be boys than girls. Interestingly, Type 4 students scored the lowest on the choice scale. This may be due to their task orientation in which they perceive the teacher as leader of instruction. Student choice on tasks would tend to delay the completion of the job.

To obtain project teachers' views of the characteristics, organization, and typical activities of their classrooms a questionnaire was developed and administered to all teachers early in the fall of the year. Each item on the instrument was selected or developed because of its relationship to factors that research had demonstrated were related to achievement in mathematics. After numerous revisions and two pilot tests, a final version was developed which contained 73 questions divided into three sections. In the first section teachers were instructed to indicate where they would classify their classrooms on a continuous scale with regard to specific classroom practices (amount of testing, emphasis on enjoyment, etc.). A second section consisted of ten items which gathered information concerning teachers' opinions, interests, and attitudes about mathematics. Teachers indicated their degree of agreement or disagreement (using a five-point scale) with a number of statements such as, "I feel I have a good, sound background in mathematics," and "Teaching math makes me feel secure." The last section asked teachers for specific quantifiable information, such as the number of days per week that they taught math. This section also included several open-ended questions that posed various instructional problems and asked teachers what they usually did to resolve such problems.

The subscales which defined the instrument and on which the teacher typologies were based are defined below.

1. *Need for Personal Control*—of classroom events and rules.
2. *Need for Contextual Stability*—in the curriculum, classroom organization, and instructional pattern.
3. *Degree of Individualization*—of children's instruction.
4. *Degree of Abstractness*—using abstract concepts, techniques, or materials with which the students have little familiarity.
5. *Degree of Security*—feeling comfortable and secure about ability to teach math.
6. *Experience*—total number of years of elementary school teaching experience plus the number of years of experience teaching mathematics.
7. *Education*—total number of credit hours in mathematics plus the number of graduate credit hours.

The derived typologies are described below and their profiles can be found in Figure 7.4 and Table 7.2.

Teacher Type 1—Less Experienced/Less Educated. Teachers who constituted this typology tended to be younger, less experienced, and have little education beyond the bachelor's degree. They desired a moderate degree of contextual stability in their classrooms and tended to present material in a nonabstract manner. Scores on need for personal control

FIGURE 7.4 Graphic Representation of the Standardized Scores for Each of the Four Teacher Types (Grade 4)

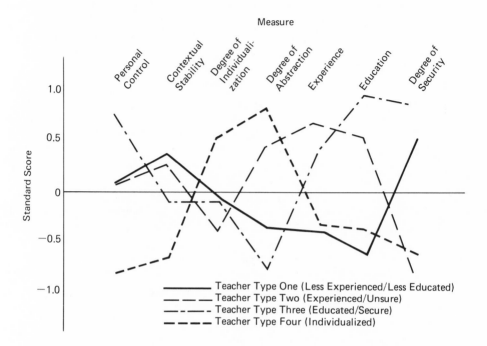

and degree of individualization were about average. They reported that they felt reasonably secure teaching mathematics.

Teacher Type 2—Experienced/Unsure. Type 2 teachers tended to have extensive teaching experience but had taken only a moderate amount of course work beyond the bachelor's level. It appears that Type 2 teachers have taken course work needed to progress on the salary schedule but few if any additional graduate hours. Teachers in Type 2 scored low on degree of individualization and degree of security in teaching math. They scored moderate on the other scales (need for personal control, need for contextual stability), with the exception of the degree of abstractness, where they indicated that they presented material in a slightly more abstract fashion.

Teacher Type 3—Educated/Secure. Four qualities (need for personal control, degree of abstractness, amount of education, and degree of security) separated Type 3 teachers from the remaining typologies. They reported a high degree of need for personal control of classroom rules, regulations, and instructional events. This feeling of personal control plus a high level of education may account for the high degree of security in

TABLE 7.2 Teacher Typologies Based on the Cluster Analysis: Means, Standard Deviations, and F Ratios for Cluster Components (Grade 4)

Teacher Typology		Need for Personal Control	Need for Contextual Stability	Components Degree of Individualization	Degree of Abstractness	Degree of Security	Experience	Education
One (N = 13)	Mean	20.00	27.53	18.53	13.61	17.84	15.30	13.15
	S.D.	2.51	5.69	6.50	2.63	1.34	7.99	5.88
Two (N = 8)	Mean	20.00	26.87	16.75	17.37	14.50	27.25	35.50
	S.D.	2.07	5.56	3.05	1.50	1.60	10.63	17.09
Three (N = 8)	Mean	22.00	24.75	18.62	12.25	19.00	23.62	42.87
	S.D.	2.61	5.25	5.47	2.65	1.41	13.19	23.64
Four (N = 10)	Mean	17.60	21.30	22.20	19.20	15.10	17.60	17.30
	S.D.	2.83	5.18	6.69	5.73	2.96	8.66	14.17
Total (N = 39)	Mean	19.79	25.23	19.12	15.53	16.69	20.05	24.89
	S.D.	2.87	5.82	5.92	4.41	2.59	10.66	19.22
F Ratio (3,35)		4.53***	2.77*	1.44	7.77***	10.89***	2.94**	8.35***

*p < 0.10. **p < 0.05. ***p < 0.01.

teaching math that this typology reported. Interestingly, Type 3 also reported the lowest degree of abstractness in their teaching approach. One might expect just the opposite, that is, the more education a teacher has the more likely he or she would be to teach from a more theoretical or abstract point of view. A rival hypothesis, however, is that because of the high education level and thus increased knowledge of the subject, the teacher might be able to dissect mathematics into easily understandable units and interlink those units into less abstract, but more meaningful lessons.

Teacher Type 4—Individualized. Type 4 teachers portrayed characteristics typically associated with individualization. They had a low need for personal control and contextual stability, and they frequently allowed students to set classroom rules, assignments, furniture placement, and the general directions of instruction. Of the four typologies, only Type 4 teachers reported any degree of individualization in their classrooms. They also reported that they teach math more abstractly, with more emphasis on theory and less on "consumer skills." The experience and educational levels of Type 4 teachers were below average, as was their security in math.

The appropriate level of mathematics subtest of the SRA achievement series served as one of the dependent variables, and the Rabinowitz-Rosenbaum Teacher Rapport Scale functioned as the attitude measure. Residual scores were calculated for the achievement measure and served as the criteria variables in an analysis of variance model with student type, teacher type, and treatment type as predictors. (See Table 7.3 and Table 7.4.) When significant interactions were found, a series of Newman-Keul multiple contrasts was performed to isolate the source of variance. Statistically significant interactions can be summarized as follows:

1. Type 1 students (dependent) did significantly better with Type 2

TABLE 7.3 Analysis of Variance, Dependent Variable—SRA Residual Score (Grade 4)

Source	df	MS	F	Probability
Treatment Condition	1	1202.90	41.16	0.0001
Student Type	3	148.84	5.09	0.0018
Teacher Type	3	313.04	10.71	0.0001
Treatment × Student Type	3	63.64	2.18	0.0876
Treatment × Teacher Type	3	129.08	4.42	0.0045
Student Type × Teacher Type	9	51.49	1.76	0.0711
Treatment × Student Type × Teacher Type	9	70.92	2.43	0.0101
Error	945	29.22		

TABLE 7.4 Number of Students, Means, and Standard Deviations of Achievement Residuals for the Various Combinations of Teacher Types and Student Types (Grade 4)

| | | | Student Type | | | | | | |
| | | 1 (Dependent) | | 2 (Independent) | | 3 (Low Achievers/Withdrawn) | | 4 (High Achievers/Task Oriented) | |
Teacher Type		Experimental	Control	Experimental	Control	Experimental	Control	Experimental	Control
1	N	30	71	13	36	38	58	5	28
Less Experience/Less Education	M	−0.72	−0.95	−0.65	−0.62	0.79	−0.93	6.35	−0.26
	SD	6.50	5.15	6.14	6.35	4.41	4.64	5.05	5.47
2	N	44	51	22	20	33	29	17	15
Experienced/Unsure	M	4.29	−1.24	3.33	−0.99	4.68	−1.58	2.95	2.32
	SD	6.72	5.20	3.84	5.57	7.46	4.11	5.37	5.21
3	N	59	11	31	7	25	21	27	4
Educated/Secure	M	2.39	−4.27	3.95	−1.65	3.46	−4.30	2.86	4.37
	SD	5.19	5.18	6.09	5.02	5.41	3.54	5.35	3.48
4	N	38	46	15	47	55	37	4	40
Individualized	M	0.17	−2.77	−6.25	−2.48	−1.00	−1.85	1.44	−1.14
	SD	5.29	5.11	4.93	5.76	5.54	4.40	6.94	5.25

(experienced/unsure) and Type 3 (educated/secure) teachers who were in the experimental treatment condition. They did significantly poorer with Type 3 (educated/secure) teachers in the control treatment.

2. Type 2 students (independent) did significantly better with Type 3 (educated/secure) teachers and significantly poorer with Type 4 (individualized) teachers, both of whom were in the experimental treatment condition.

3. Type 3 students (low achievers) did significantly better with Type 2 (experienced/unsure) and Type 3 (educated/secure) teachers in the experimental treatment and poorest with Type 3 (educated/secure) in the control.

4. Type 4 teachers (individualized) did worst with Type 2 student (independent) in the experimental treatment condition. Individualized teachers did not do significantly better with any student type under either the treatment or control condition.

5. Type 3 teachers (educated/secure) did significantly better with Type 4 students (high achievers) in the control but poorly with Types 1 (dependent) and 3 (low achievers), both in the control condition.

6. Type 1 students (dependent) who were in the experimental treatment did best with teacher Type 2 (experienced/unsure) and worst with teacher Type 1 (less experienced/less educated).

7. Type 2 students (independent) and Type 3 students (low achievers) who were in the experimental treatment did significantly better with Teacher Types 3 (educated/secure) and 2 (experienced/unsure). Independent students did poorly, on the other hand, with Teacher Types 1 (less experienced/less educated) and 4 (individualized).

8. Type 4 students (high achievers) did not do significantly better under any teacher type.

As we have discussed elsewhere, the active teaching model used in this study proved effective and substantiated earlier reviews of correlational research and the findings of others. Importantly, the interactions among student type, teacher type, and treatment type also produced findings which allowed a meaningful interpretation. For instance, Type 3 teachers (educated/secure) were quite unsuccessful with dependent and low achieving students who were in the control treatment. On the other hand, they did best with high-achieving students in the control group. Type 3 teachers are likely to be subject orient and to place more emphasis on accomplishment than on social concerns; therefore, this finding seems reasonable.

Low-achieving students are likely to get lost in an environment that stresses achievement without lesson structure. Dependent students need more attention and specification than they are likely to get in an achievement-oriented environment. Interestingly, as one would predict, low-achievement and dependent students in the experimental treatment do significantly better than those in the control group. This is possibly due to the model's emphasis on review (which increases exposure to content), homework (needed practice), and structure (which measures to dependent children to some extent). Also, the increased attention on development and the meaning of the mathematics being presented may have allowed these students to better understand their task and to be more secure in their role as students.

Type 4 teachers (individualized) were relatively ineffective with all student types and especially with Student Type 2 (independent) in the experimental treatment and, to a lesser extent, Type 2 students in the control. Although this finding may seem contrary to "conventional wisdom," it is not. Student Type 2 was labeled as behaviorally independent, but it does not follow that independent students can effectively structure their time to meet goals set by the teacher. They are just as likely to direct their attention and efforts to non teacher sanctioned events. A clear indication of this possible nonacademic orientation was the relatively high incidence of behavioral problems reported for independent students. Type 4 (individualized) teachers may have allowed too much latitude in choice, and it may have become dysfunctional, especially for students who were likely to "take advantage" of the situation.

Type 1 students (dependent) did best with Type 2 teachers (experienced/unsure) in the experimental treatment condition and, second best with Type 3 teachers (educated/secure), also in the experimental treatment condition. On the other hand, they did poorest with Type 3 teachers (educated/secure) in the control group. The relationship between Student Type 2 and Teacher Type 3 has been discussed previously; however, the discovery that experienced/unsure teachers in the experimental treatment did best with dependent students is new. This finding seems to add support (at least in the short run) to those who advocate matching students and teachers on the basis of the similarity of their characteristics (see Cronbach and Snow, 1977). It seems natural that a symbiotic relationship would have developed between a teacher who lacked security in teaching math and a student who needed teacher support.

Unsure teachers also seemed to benefit extensively from the experimental treatment program. Apparently, the structured program provided these teachers with the direction they lacked and for which they were perhaps searching. Interestingly, experienced/unsure teachers did consistently better in the experimental treatment, regardless of the student type they taught.

Type 2 students (independent) did best with Type 3 teachers (educated/secure) in the experimental treatment; whereas, they did very poorly with Type 4 (individualized) students in (the experimental treatment) condition. The rationale for the poor performance of Type 2 students (independent) when paired with Type 4 teachers (individualized) in the experimental treatment seems to stem from a lack of student academic involvement, due to relatively poor implementation of the program by these teachers.

The explanation of why independent students did best with educated/secure teachers in the experimental treatment seems to correspond nicely with the academic involvement hypothesis suggested above. Since independent students are less conforming to teacher expectations and sanctions, it is clear that they need firm encouragement to stay on task as defined by the teacher. Clearly, Teacher Type 3 (educated/secure) would be likely to provide such task direction and, therefore, would do better with independent students. The experimental treatment serves to further enhance this task emphasis.

The last teacher type by treatment type interaction that was statistically significant occurred with Type 3 students (low achievers). Students in Type 3 (low achievers) did substantially better with Teacher Types 2 (experienced/unsure) and 3 (educated/secure) in the experimental treatment. Conversely, they did the poorest with Type 3 teachers in the control treatment. The relationship between Teacher Type 3 and Student Type 3 has been discussed previously. The new finding that low achievers did best with Type 2 teachers (experienced/unsure) in the experimental treatment again seems to support the idea of matching (at least in the short run) student-teacher characteristics for optimal growth. In this case, the matched characteristic would be insecurity in mathematics. However, simple matching does not entirely portray the picture; otherwise, Teacher Type 2 (experienced/unsure) in the control condition would also produce large student gains. The fact that little gain resulted indicates the mediating effect of the experimental treatment. Seemingly, the increased structure and enhanced attention to meaning and development afforded by the experimental treatment have a positive effect on both the teacher and the student. The student probably benefits most from the increased practice and review, while the teacher benefits from the increased direction that comes from greater organization.

One of the more interesting findings of this study was the interactions between teacher type and treatment type. There exists a strong teacher effect in the treatment condition that is not found in the control sample. This interaction occurs for Types 2 (experienced/unsure) and 3 (educated/secure) teachers but not for Teacher Types 1 and 4. An examination of the mean implementation scores for the teacher types in the treatment group revealed that Teacher Types 2 and 3 implemented significantly more of the treatment behaviors than did Teacher Types 1 and 4 (Means: Type 1 = 8.48, Type 2 = 9.82, Type 3 = 9.64, Type 4 = 8.25). The data collectively suggest that teachers who implement the model get good results, yet some teacher types choose to use more facets of the model than other teacher types.

Since people will more likely adopt and internalize ideas that are consonant with their existing beliefs, one could predict that teachers who already believed in an active instructional model or teachers who were unsure using their present instructional strategies would be most likely to implement the experimental treatment program if requested to do so. For example, Type 3 teachers (educated/secure), who indicate they teach in a more direct manner, would be more likely to employ the experimental treatment program than Type 4 teachers, who prefer to teach using an individualized model. Similarly, Type 2 teachers (experienced/unsure) would probably enhance the treatment because it resembles the "old" method of teaching with which they are familiar, and because they indicate they are currently insecure teaching math in the present manner. Teacher Type 1 (less experienced/less educated), on the other hand, showed a high degree of security in teaching in the present manner, and therefore would not be likely to change without additional and more specific training in how to change.

Although the small sample size in some cells and the employment of self-report measures demand that considerable caution be exercised in any attempt to generalize the results of this particular segment of the larger study, the overall results do offer convincing evidence that interactions between and among student types, teacher types, and treatment types exert important influences on students' mathematics achievement. We do not mean to imply, however that all effective instructional techniques are situational—quite the contrary. The active teaching model as operationalized by the treatment program proved quite effective in most instances. What we are suggesting is that modifications in the active teaching model may increase its effectiveness in some defined areas. For example, for those teachers who were uncomfortable with the development request because of lack of a sufficient mathematics background or poor delivery techniques, it would be necessary to alter the development portion of the lesson such that the goals are still accomplished but through a different vehicle. We have tried to make the development portion of the lesson clearer for such teachers, but other teachers may need increased mathematical understanding before they can improve the development part of their lessons (see Chapter 9).

A different form of program adaptation can be found in the work of Janicki and Peterson (1981). They exposed all students to our basic active instructional model (i.e., teacher presented the lesson actively) but modified the model to include a small-group variation in the controlled practice portion of the lesson for some students. Results from their study indicated that certain students who received the small-group variation outperformed their counterparts in the study who did seatwork independently. Part of the reason that the *interaction effects* between teachers, students, and the program have been relatively small (in terms of immediate, practical implications) may be due to the fact that various program aspects have different effects upon different types of students (homework is useful to some but irrelevant to others) such that the particular advantages and disadvantages of the program for certain types of students average out and are masked by the general effectiveness of the program per se for all students.

Summary

We have attempted to define and conduct research on teacher and learner styles. We have reviewed the literature and have presented a conceptual framework for organizing the research that has been completed in this area. We have discussed some of the teaching and learning style research that we have conducted within the Missouri Mathematics Program. We have seen that when the program is well implemented the learning gains of students in the experimental group exceed those in regular control classrooms. However, the data do suggest that certain types of teachers implement more of the program than do others and future research and development are needed if we are to understand why this occurs and to determine if certain teachers can be encouraged to use (appropriately) more aspects of the program. In particular, it appears important to conduct more research such as Janicki and Peterson's (1981), in which specific aspects of the program are modified in order to determine how such changes influence particular combinations of teachers and students.

8

Teaching and Learning Styles
and the Missouri Mathematics
Program

In the previous chapter we provided an overview of several studies that have focused on teacher and student typologies and the interactions among them. In addition, a schema was presented to aid in understanding the various traits and the methods that have been employed to identify typologies. Finally, we examined our work at the fourth-grade level and discussed some of the implications of differential program effects. The purpose of this chapter is to extend the discussion of the interactions among teacher and student typologies to the sixth and eighth grades. Specifically, we will discuss three areas: the similarities and differences that were apparent across the fourth-, sixth-, and eighth-grade student and teacher typologies; the general interaction patterns we discovered; and the implications of this work for altering the active teaching model. However, prior to an examination of these substantive issues we need to consider one important procedural detail: the stability of aptitude measures.

APTITUDE STABILITY

The number of ATI investigations has increased markedly recently, in part because of the belief that learning can be improved when students with particular types or levels of aptitudes are placed into treatments compatible with those aptitudes. Largely as a result of Bracht's (1970) review of ATI studies (which suggested a lack of significant ATI when cognitive aptitudes were used), and Cronbach and Snow's (1977) definition of aptitudes as "any characteristic of a person that forecasts his probability of success under a given treatment" (p. 6), several recent studies have used noncognitive variables as aptitudes within the ATI model (Ebmeier and Good, 1979; Peterson, 1977; Cunningham, 1975; Bennett, 1976; Solomon and Kendall, 1976).

The use of noncognitive aptitudes, however, presents an interesting problem within the context of the ATI model. One assumption underlying the model is that aptitudes remain fairly stable over time (at least over the duration of the treatment). If the aptitudes are not sufficiently stable, then as students change with respect to these aptitudes they may become "mismatched" with respect to a particular instructional treatment.

While the stability of most cognitive aptitudes has been well established over the years, the stability of most noncognitive aptitudes is largely unknown. Anderson and Liu (1979) have addressed this issue. They administered two aptitude instruments (Locus of Control and Student Preference Report: Form A) to 75 sixth-grade urban school children in September and again in May. Although they concluded that aptitudes and, therefore, typology classifications fluctuated considerably, their conclusions can be questioned on methodological grounds. They classified students on a mean split basis instead of an extreme group basis, thereby insuring that some members of one classification were in fact closer to some members of the other group than they were to members of their own typology. Considering the error variance present in most aptitude measures and the mean split clustering technique used in this study, the changing classification over time of 34 percent of the students in the sample should not be unexpected. Still, the study by Anderson and Liu raises an important issue and motivated us to examine the stability of our affective measures.

Data from our fourth-, sixth-, and eighth-grade experimental studies shed some additional light on the stability question. We found a moderate degree of stability between the September and January administrations of aptitude measures employed in our studies. The correlations between the scores from the pre- and postadministration were all above 0.28 and were all significant at the .01 level. Part of this stability may be due to the fact that since we were aware of the possibility of short-term instrument instability, we had taken care in the development of the aptitude inventory to include only questions which fourth-grade students from pilot studies had shown some consistency in answering over a two-week time interval. Furthermore, our two time measurements were closer together than those in the Anderson and Liu study.

Some of the instability can naturally be attributed to teacher and treatment effects, although at this point we do not know how much. Just as we observed a change in student achievement, one would expect similar changes in students' beliefs and preferences (e.g., conscientiousness).

Indeed, research that explores how various combinations of student, teacher, and program characteristics affect the noncognitive aptitudes of students might be very productive because it deals directly with variables that influence learning. For instance, if we know that a particular pattern of teacher behaviors in a defined context results in increased conscientiousness in low-achieving students, then we could utilize this knowledge to increase the frequency of homework completion, which ultimately should result in better math achievement. In many ways this type of investigation parallels the time-on-task research which is typified by a movement away from the exclusive study of outcome variables (see Anderson, 1981, for an example of work in this area).

CHANGES IN STUDENT PREFERENCES OVER TIME

We have seen that the measures we used to study student preferences had some stability. We now turn to a consideration of the stability of students' preferences across grade levels. Table 8.1 indicates significant directional shifts ($p.<01$) in the response patterns of the fourth-, sixth-, and eighth-grade students on four of the seven scales of the aptitude inventory. These changes are important because they suggest that some changes in the Missouri Mathematics Program might help to improve the program at these grade levels. It should be stressed, however, that although students at the various grade levels had statistically different scores on four of the seven dimensions, the magnitude of some of these differences was not large and many may not be of any educational significance. At this point in our investigations we are unable to offer any definitive advice but can only suggest that some of these differences might indicate important variables to study in future research.

One of the changes we noticed over grade levels was that older students indicated that they required less external motivation to complete tasks and were less enthusiastic about doing mental computations than

TABLE 8.1   Mean Scores on the Student Attitude Inventory Across Grade Levels

	Grade Level		
Component	*4*	*6*	*8*
Mental Computations	2.46	2.31	2.09
Conscientiousness	7.32	7.50	7.29
Choice	1.11	1.64	1.79
Dependence	4.44	4.42	4.46
Other Orientation	1.87	2.13	2.16
External Motivation	4.55	3.92	3.69
Behavior	0.50	0.49	0.53

their younger counterparts. Younger students, on the other hand, preferred less choice and indicated less desire to work with other students. These trends may reflect both a response bias difference among the student populations based on maturation differences as well as context differences among the three grade levels in mathematics. As the complexity of mathematics increases with grade level, teachers are compelled to alter their instructional strategies, and this undoubtedly plays an important role in the development of student preferences. For instance, because of the difficulty of mentally manipulating algebraic symbols, eighth-grade teachers may not ask their students to solve as many mental computation problems as fourth-grade teachers who are focusing on multiplication and division facts and computation with multidigit numbers. Because of this difference in both academic content and appropriate practice, mental computation must be carefully planned and executed, especially at the upper-grade levels. Without conscientious planning by the teacher, having students work problems in their heads is not likely to be productive. Hence, one area of promising research is to vary the way in which the mental computation phase of a lesson is conducted in secondary classrooms and to assess the influence of different treatments on particular students.

A second change in student aptitude over grade levels involved the amount of external motivation needed for homework assignment completion (fourth grade $x = 4.55$, sixth grade $x = 3.92$, and eighth grade $x = 3.69$). Since the amount of homework assigned increases dramatically over grade levels, upper-level students are much more likely to expect regular homework. In contrast, homework assignments at the lower grade levels are more sporadic (assignment of homework tends to be more random). Considering the increasing maturity of students over grades and the increased pedictability of teacher-assigned homework, it is reasonable to expect that younger students will feel as though they need more outside encouragement. If the predictability hypothesis is correct, then the most explicit directions and expectations related to homework should occur at the lower grades.

In addition, because of the structure of most elementary classrooms, where there are no distinct indicators such as bells that signify a change in task, most teachers need to provide anticipatory cues that tell students that a new activity is beginning and link the new activity to previously learned material (hence, the importance of smooth transitions between lessons may also be more important in program use at the elementary school level than in secondary classes).

Two other changes which occurred across grade levels (see Table 8.1) were the preferences of older students for more choice and for greater freedom to work with other students or adults. Social maturity may account for these reported changes. Since older students have a larger experiential base, it is only natural that they should prefer more contact with other individuals and more variety in tasks and assignments. A more subtle factor that would account for this trend might simply be the increased difficulty of mathematics at the upper-grade levels. As content increases in difficulty, fewer students are able to understand it easily. Students can cope with (or "adjust to") this problem by selecting less challenging math problems (if given the choice) and seeking the assistance of other students and adults.

Student choice can be used by teachers to motivate students; however, it can also serve as an escape mechanism for some students. (This is why no simple relationship exists between student preference for a learning format and student achievement.) Allowing students options has long been recognized as an important aspect of motivation. The implication is that as students become older, teachers can allow them more choices, thereby increasing student motivation. The point is not that teachers should indiscriminately increase the number of options as students grow older; rather, they should offer alternatives which interest students relative to previous learning, and options which assist in accomplishing instructional objectives (e.g., more opportunity to work in small peer groups for review). However, from our observations, just the *opposite* occurs, because students are often afforded more choices in selection of content and learning format at the lower grade levels. Such options gradually decrease as students progress through the grades, typically culminating with almost no options in some graduate level college courses.

The second way students can minimize the problems inherent in more difficult course content is to seek the assistance of more capable students and/or adults. This communal working of problems affords the slower student two advantages. First, if a student wants to know how to solve a problem but is unable to do so, the assistance provided by another student can supplement the teacher's instruction and provide some guided or corrective practice. In addition, a student's peers are often able to explain a problem's solution in a way different from the teacher, so that the student no longer is confused. On the other hand, if a student wants to avoid working math problems, the opportunity to work jointly with others is an excellent opportunity to rely on someone else to complete the work and thus avoid the adverse effects of not having the assignment completed.

The above rationale applies to students located in the middle of the ability distribution, but tends to break down sometime between the sixth and eighth grades for very poor and very good students. That is, eighth-grade low-achieving students actually indicated less desire ($x = 1.29$) to seek help from classmates than did their counterparts in the sixth grade ($x = 1.63$), and very high-achieving eighth-grade students ($x = 2.71$) showed slightly less inclination to help other students than did sixth-grade high achievers ($x = 2.76$).[1] Because of the more structured environment of the eighth grade, the increased peer pressure for conformity, and the diffi-

---

1. Although the actual difference in means between the sixth- and eighth- grade achievers on the other orientation scale is small, the eighth-grade scores are much lower than what one would predict if extrapolating from the fourth- and sixth-grade data.

culty of the mathematics content, this trend is understandable. Low-achieving students at the eighth grade are likely to achieve several years below grade level. In effect, a few are almost beyond the help of their classmates. This factor, likely coupled with intense peer pressure to avoid looking "stupid," results in a withdrawal defense mechanism.

High-achieving sixth-grade students probably are rewarded by teachers for helping other students, resulting in the high-reported desire of sixth-grade students to work with others. By the eighth grade, however, teachers place more emphasis on academic achievement, and high-achieving students are thus likely to perceive that any interaction with students of lesser ability is nonproductive and will only delay the completion of teacher assigned tasks. Peterson, Janicki, and Swing (1981) reached a similar conclusion when teaching a geometry unit to mixed ability fourth- and fifth-grade students. Although their work did not include comparisons across grade levels, they noticed that when high-ability students were in a setting that required them to interact more frequently with low achievers (e.g., peer tutoring), the high achievers had significantly lower attitudes toward that teaching approach than when they were on their own. The desirability of decreasing contact between high and low achievers is dependent on the teacher's objectives. However, if little contact occurs, teachers will find it difficult to get high achievers to help low achievers. Furthermore, older low-achieving students are less likely to approach the teacher or other students for assistance. Accordingly, teachers need to initiate more contacts with these students in order to maintain involvement levels and influence learning. The fact that eighth-grade low and high achievers prefer less contact with other students also has implications for teachers who utilize peer tutoring or heterogeneous group work. Clearly, to promote positive interaction, especially in mathematics, teachers will need to restructure *reward systems* so that high-achieving students see a reason to spend their time helping slower students and low achievers will not feel threatened by such involvement with more able students.

Fortunately, there is work to suggest that classroom reward systems can be shifted (Slavin, 1982; Bossert, 1979), and other studies indicate that teachers can encourage students to work together successfully in learning teams (e.g., Webb, 1982; Peterson and Wilkinson, 1982), at least for a short time. Such data should encourage teachers to consider encouraging greater student involvement and choice in secondary mathematics classes (especially in controlled practice and review/maintenance activities).

STABILITY OF STUDENT TYPOLOGIES

We have noted that students' preferences within a given year show moderate stability and we have discussed some of the differences (and possible implications) across grade levels. The reader will recall that in the previous chapter we reported our attempt to categorize students on the basis of their responses to the aptitude inventory. It is also important to examine the stability of student typologies across grade levels. The descriptions of the fourth-grade data appear in Chapter 7.

Although the potential for different student typologies across the three grade levels was high, we found that the three sets of typologies were remarkably similar. For instance, when the within-grade rankings on the aptitude subscales are compared to the rankings of similar typologies at the other grade levels, few differences exist. Of 84 possible rank comparisons (e.g., comparing the rank of high achievers at the fourth grade with the rank of sixth- and eighth-grade high achievers on the *other* orientation scale), 42 were identical and only 12 varied by more than one rank. This stability partly reflects the high correlations among the subscales and preachievement measures, correlations which led us not to expect the paired typologies to differ to a large degree. High-achieving students are always likely to score high on conscientiousness and low in reported behavioral problems, irrespective of the grade level or subject area under study.

When the student typologies resulting from our work in the fourth, sixth, and eighth grades are compared to student typologies at similar grade levels devised by others, there is again substantial overlap. This is especially true for typologies describing low and high achievers, which are included in almost all studies. In addition, the descriptions of these classifications tend to be more concise and homogeneous than those representing students in the middle of the achievement distribution. This lack of clarity in the literature about typologies representing more average or typical students seemingly stems from two sources. First, there is less conceptual agreement among authors concerning important traits that describe these students. Secondly, even after the profiles or characteristics have been identified, the high degree of variance in the questionnaire responses of average achievers results in typologies with some degree of overlap.

Although there are no unique student types, because all students share characteristics descriptive of all the typologies at least to some degree, we have clustered the student typologies suggested by the numerous studies into four basic divisions (see Table 8.2). Clusters one and four represent high- and low-achieving students respectively, and are reasonably self-evident. Cluster two students probably would be categorized by teachers as good students who are hard workers. Although these pupils are not the most able students in the class, through persistence they perform at relatively high levels. Over-achievers would most likely be found in this category. Students in cluster three tend to produce average work, but only if they are carefully monitored and encouraged. Without that attention, it is unlikely that they will perform acceptably.

TABLE 8.2   Classification of Student Typologies from a Number of Studies

Characteristics	Studies	Label
Cluster One		
High Ability	Whitzel & Winnie (1976)	High Achiever/High Ability
Very High Achievement	Peterson (1976)	Independent/High Achiever
Highly Motivated	Bennett (1976)	Type 8
Confident	Solomon & Kendall (1976)	Type II (Prior Achievers)
	Good & Power (1976)	Success
Social	Present Study 4th Grade	Type IV (High Achievers)
Low on Self-direction	Present Study 6th Grade	Type IV (High Achievers)
Few Behavioral Problems	Present Study 8th Grade	Type IV (High Achievers)
Cluster Two		
Moderate Achievement	Solomon & Kendall (1976)	Type III (Moderate Achievers)
Social	Good & Power (1976)	Social Student
Dependent on Teacher	Bennett (1976)	Type IV
For Some Structure	Peterson (1976)	Low Ability, Conforming
Moderately Low Need for	Present Study 4th Grade	Type I (Dependent)
External Encouragement	Present Study 6th Grade	Type II (Average)
	Present Study 8th Grade	Type I (Moderate Achievers)
Cluster Three		
Low Achievers	Good & Power (1976)	Phantom Students
Dislike Public Exposure	Bennett (1976)	Type 7
Not Dependable	Peterson (1976)	Independent, Quiet
Somewhat Withdrawn	Present Study 4th Grade	Type II (Independent)
Capable of Achievement	Present Study 6th Grade	Type I (Low Achievers)
If Given Assistance	Present Study 8th Grade	Type III (Low Achievers)
Cluster Four		
Very Low Ability	Whitzel & Winne (1976)	Low Achiever
Sometimes Desire Public	Peterson (1976)	Low Achiever, Low Anxiety
Attention but for	Bennett (1976)	Type 3
Non-academic Reasons	Solomon & Kendall (1976)	Type I (Low Achievers)
Desire a Lot of Choice	Good & Powers (1976)	Alienated
Very Withdrawn	Present Study 4th Grade	Type III (Low Achievers)
Need Constant		
Encouragement	Present Study 6th Grade	Type III (Withdrawn/Very Low Achievers)
Can Cause Behavioral		
Problems	Present Study 8th Grade	Type II (Withdrawn/Very Low Achievers)

TEACHER BELIEFS AND PREFERENCES

We wanted to examine the possibility that the Missouri Mathematics Program would need to be modified for teachers across the three grade levels to retain its general effectiveness. Thus, we administered the Teaching Style Inventory (see Appendix A) at all three grade levels to obtain teachers' views and preferences about mathematics instruction.

From examination of responses from fourth-, sixth-, and eighth-grade teachers (see Table 8.3) we noticed several differences among teachers at the three grade levels. The most obvious difference was an increased teacher directiveness in the eighth grade. These teachers more frequently expressed a desire to be in control of classroom events and to limit student options than did sixth-grade teachers, who in turn wanted more control than the fourth-grade teachers (recall that older students are expressing more interest in choice!). Similarly, eighth-grade teachers reported less frequent use of techniques commonly associated with individualized instruction (e.g., diverse lesson plans) and were more likely to teach their classes as a whole, give common assignments, and hold similar expectations for all students. Indeed, with the exception of the personal control dimension (where sixth- and eighth-grade teachers were more alike than sixth- and fourth-grade teachers), sixth- and fourth-grade teachers usually reported using similar instructional techniques.

A second grade-level difference among teachers was that eighth-grade teachers reported feeling more secure in teaching mathematics than their counterparts at lower grades. Most of these teachers were better trained to teach mathematics (more stringent state certification requirements) than the lower-grade teachers and, in addition, the departmental nature of junior high schools allows secondary teachers to specialize.

The degree to which teachers presented material in an abstract manner increased consistently from the fourth to the eighth grades. This means that as students progress through grade levels they or their teachers were less likely to intentionally relate curriculum content to students' previous experiences. Eighth-grade pupils would thus more frequently encounter assignments containing symbolic, theoretical problems as opposed to concrete, practical problems, which would be presented to younger students.

Although these differences in reported instructional patterns are important and have practical significance, they are not surprising. What the data do vividly point out is the increasing trend toward more teacher-

TABLE 8.3   Mean Scores on the Teaching Style Inventory Across Grade Levels

	Grade Level		
Component	*4*	*6*	*8*
Personal Control	19.79	32.40	34.00
Context Stability	25.23	26.40	32.20
Individualization	19.12	18.93	12.40
Degree of Abstraction	15.53	16.86	17.53
Degree of Security	16.69	16.86	17.73
Experience	20.05	19.93	23.20
Education	24.89	46.20	55.66

directed, controlled, and structured mathematics instruction with increasing grade levels. There are probably several reasons for this trend, some of which are related to the structure of mathematics as an academic discipline. Increased teacher directiveness also depends upon teachers' role expectations and their training. For instance, abstraction level of mathematics content and the concomitant pressure to cover a defined (by the expectations of the next course) amount of material probably cause most upper-grade-level teachers to believe that a more direct approach to instruction and control of extraneous variables are the most efficient ways to convey a maximum amount of material in a limited time.

In addition, the marked differences between eighth-grade teachers' lessons and those of the elementary school teachers could be logically related to the role models these teachers observed while they were in college. Because eighth-grade teachers must have at least a minor in mathematics to be certified, most, if not all, of their models of mathematics teaching have been faculty with appointments in mathematics departments of colleges or universities. In college courses taught by mathematics instructors, lecturing is the predominant mode of instruction. It is therefore not surprising that secondary teachers employ many of the same instructional patterns they observed as college students. In contrast, elementary teachers are required to take few mathematics courses outside the methods courses within the college of education, and therefore do not experience the same role modeling as secondary teachers. This difference in experience could be manifested in several ways, one of which has great significance for educational programs which require active teaching. For example, it is difficult for elementary teachers to devote 15 minutes to developmental activities which focus on the meaning of the mathematical ideas being studied (a reasonable expectation in a direct instruction setting). Indeed, many elementary teachers are likely to encounter difficulties even finding enough substantive mathematics material to present in daily direct lessons, because most elementary math curricula are designed to produce speed and proficiency in basic math skills, not conceptual understanding. Also, because of their inadequate backgrounds in formal math and their lack of direct instruction role models, many elementary teachers probably lack sufficient prerequisite knowledge of mathematics or instructional skills necessary to consistently deliver 15-minute developmental lessons. From our observational work over the last few years it is apparent that little developmental instruction occurs in elementary mathematics classes (others have made similar observations in reading, see e.g., Durkin, 1980). In contrast, most junior high school teachers typically include in their lessons a formal, conceptual development component whose characteristics closely parallel the key attributes suggested by the active teaching model. Hence, implementing the development portion of an active teaching treatment program should be immensely easier in the upper grades, because teachers are familiar with many of the suggested teaching behaviors and possess the required mathematical knowledge. Indeed, while increasing active teaching might be the most perplexing problem in elementary mathematics instruction, finding a control group for related research is likely to be the most persistent problem at the upper-grade levels. However, as we have noted previously, the *quality* of the development lesson is the issue in secondary schools (see Chapter 10 for an extended discussion of the development part of the lesson).

Finally, the reported tendency of upper-grade teachers to be more abstract in their instruction has implications for lesson design. For example, it is commonly recommended that a well-designed mathematics lesson should progress from the familiar to the unknown and from the concrete to the abstract in order to increase motivation and enhance transfer effects between lessons. If upper-grade teachers do utilize more abstract material by including fewer familiar and concrete examples, as indicated by their responses, then they must be very careful in their choice of examples and in the way examples are incorporated into development. This task is made especially difficult by the reliance on previously learned material and the lack of readily usable, concrete, or practical examples. Much eighth-grade mathematics content (e.g., algebraic equations or square roots) has eclipsed that which students normally use on a daily basis, and the teacher must therefore rely on more abstract examples for illustrations. This problem is further compounded by the fact that many secondary students over the years have not maintained an adequate pace, and therefore many teacher examples and referents to previous material may be of little utility. In contrast, lower-level teachers can more easily construct practical examples, since elementary mathematics content basically focuses on skills commonly used on a regular basis by the majority of adults.

Program developers who wish to employ an active teaching model must therefore make certain that both the teacher training components and instructional recommendations are appropriate for various grade levels. The development portion of lessons at the lower-grade levels needs to be somewhat shorter, but the teacher training component must be stronger than at the secondary level. Many teachers at the lower levels need increased knowledge of mathematics so that they are able to teach the conceptual development segments. These teachers should also practice active teaching techniques until they can easily apply them. Secondary programs should focus more on the refinement of teachers' existing skills for teaching development. Considering that many secondary teachers employ many instructional behaviors advocated by the active teaching model, it should not take long for them to fully utilize this model. Although active teaching or direct instruction may be more prevalent in secondary schools, the quality of active teaching is still an issue (see Chapter 10). Development seems critical at the secondary level because of the complexity of the mathematics and this segment probably needs the most careful attention. It seems profitable to engage in lengthy discussions with secondary teachers in order to use their expertise when developing programs (see Chapter 6).

TEACHER TYPOLOGY COMPARISONS

Because of our relative success in clustering students into typologies that were similar across the three grade levels, it seemed possible that a parallel situation might exist with the previously derived teacher typologies. Athough we isolated four distinct patterns of teacher behaviors and found teachers at each grade level who exemplified these broad classifications, teacher types were not as easily matched as student types. When the student typologies were condensed, only 12 of the possible 84 rank comparisons differed by more than two ranks (e.g., comparing fourth-grade high-achievers' rank on the conscientiousness scale with eighth-grade higher-achievers' typologies ranking). In contrast, there were 15 mismatches in teacher profiles and only 34 cases where the rankings were identical. This indicates that teacher, classroom, and instructional characteristics which are associated at one grade level are not necessarily related to one another at other grade levels.

Despite the problems discussed above, was generated four generic descriptions of teacher types present in fourth-, sixth- and eighth-grade classrooms. One of the four descriptions of characteristics that fit together we labeled as *individualized*. These teachers had little need for classroom stability and exhibited a good deal of diversity in instructional patterns, assignments, classroom rules, and the expectations they held for students. The second broad typology represented teachers at all three grade levels who felt somewhat *insecure* teaching mathematics. They reported great uncertainty in knowing how to organize and deliver a mathematics lesson, as evidenced by contradictory responses to the teaching style inventory. The third typology present in all three grade levels we labeled *secure/educated/experienced*. Although there was more internal variation among members of this group than the previous two broad categories, in general they were experienced, highly educated individuals who maintained a moderately high degree of classroom control with little individualization. A fourth cluster of teachers, *secure/young/ direct*, were described by their rigid control of classroom structure, instructional time, lesson assignments, and student behavior. They seemed reasonably secure in teaching mathematics, were more concrete in their lessons, and tended to be less experienced.

TEACHERS AS DECISION MAKERS

The formation of student and teacher typologies does not, of course, lead to direct changes in the Missouri Mathematics Program (or any program) but it does provide a way for thinking about program adaptation. If students at a particular grade level generally prefer identifiable aspects of a teaching program or if typologies of students (and/or teachers) have distinct needs and preferences then it is possible to attempt to relate specific changes in the mathematics program (i.e., related to identified needs) to changes in classroom behavior as well as achievement and attitude. Obviously, changes in the program (e.g., building in more choice for older students) may not improve student achievement and indeed, program adaptation may even lower achievement.

In review, we have built a mathematics teaching program and we have tested the program as a unit (teachers were encouraged to use *all* aspects of the program and even the degree of emphasis of lesson parts—how much time was spent in development has been controlled to some extent). Thus we know that the program positively influences student achievement.

We believed at the time we developed the program that it would be effective in producing gains in specific contexts if used as a system of instruction. For example, we felt then, and still do, that in most mathematics lessons there should be a brief review of assigned work and concepts, active conceptualization of the meaning of the mathematical ideas being studied, a chance for students to practice and apply knowledge, and opportunities for teachers to assess immediately whether or not students have comprehended the lesson. However, we realize that some lessons may require only ten minutes of development, whereas others demand most of the available class time.

Time needed for development varies in part with the concept being taught, stage of lesson development (first time concept being taught vs. review), the quality of the teaching (usefulness of examples chosen, clarity of instruction), and the types of learners. Hence, the amount of time needed for development varies with the particulars of a given classroom. Although it is possible to state general guidelines (most mathematics lessons need a development phase) and to suggest certain minimum times (ten minutes of development is suggested for any lesson), individual teachers ultimately have to make decisions about time allocations and general quality and format (Will students work in small groups or alone during controlled practice? Will teachers allow students to do some of the development themselves?).

As Slavin (1982) has noted, too often instructional methods are developed and evaluated and remain in a state of "premature closure" (the program continues to be used in the same way). We certainly do not advocate that our program be used in only one way. Indeed, our examination of the stability of teacher and student types and the implications of students' and teachers' beliefs and preferences is directed toward consideration of, and recommendations for, program adaptation. However, we must emphasize that the program was tested as a whole and in a reasonably standardized way and that program users who vary certain aspects of the program for all students or parts of the program for certain types of students need to evaluate the effects of such changes in their own settings.

TEACHER AND STUDENT TYPOLOGY INTERACTIONS

We have discussed teachers' and students' beliefs and preferences, both from the standpoint of typologies (Do some groups of teachers and students have sanction beliefs and/or needs?) and in terms of changes in responses over grade levels.

In Chapter 7 we presented a detailed examination of the interactions among the active teaching model, teacher type, and student type at the fourth-grade level. We did this to familiarize the reader with the general methods which we used and to illustrate how one might actively investigate this area. Rather than present the details of interactions at the sixth- and eighth-grade levels, in the following section we will discuss only major and consistent findings from the Missouri Mathematics Project and integrate the observations of other researchers where appropriate (see Ebmeier and Good, 1979; Good and Grouws, 1975, 1979, 1981; and Beckerman, 1981 for a more comprehensive treatment).

Considering the number of interactions among the three grade levels on the two dependent variables (attitude and achievement), even to present the major findings is a complex task. In addition, to integrate the work of our colleagues who employed different designs across different grade levels, and in some cases even used students from different countries, is truly a formidable task. The conclusions discussed below would obviously have been stronger if the studies had more closely overlapped; however, such correspondence between studies is typically not found in the social sciences. As a result, although there is a reasonable amount of support for the following conclusions, they should be only tentatively accepted at this point.

One of our most interesting findings was that students who scored low on the dependence scale and otherwise displayed characteristics of independence did relatively poorly academically in nonstructured environments. Similar results were reported by Whitzel and Winne (1976), who concluded that low achievers with high field independence had greater mathematics achievement gains in classrooms that were more structured; by Peterson (1977), who reported greater achievement gains for low-conformance students in high structure/low participation settings; Bennett (1976), who found that independent, motivated, and extroverted students performed better in formal classrooms; and by Solomon and Kendall (1976), who suggested that children characterized as social, independent, motivated, and direct (but not academically oriented directions) did better in traditional classes. Although this finding may seem contrary to conventional wisdom, which suggests that independent students would do better in individualized environments, it is not. Simply because students are labeled as behaviorally independent does not mean that they can effectively structure their time to meet goals set by teachers. Such students are just as likely to direct their attention and efforts to nonacademic activities. A clear indication of this possible nonacademic orientation is the relatively high number of behavioral problems reported by independent students. Individualized settings may allow these pupils too much choice. In these situations, choice may become dysfunctional, especially for students who are likely to take advantage of their freedom. In contrast, more structured environments tend to keep independent students actively involved in academic tasks and hold them more accountable (e.g., the students know their work will be periodically checked). Increased structure may thereby provide more academic successes and motivate students to continue working on teacher-directed tasks. Another explanation for independent students' better achievement in structured classes might be related to the more frequent and more public teacher-student interaction in structured classrooms, as opposed to more isolated environments occurring at the learning stations typically found in individualized classrooms. Considering independent students' propensity to explore new situations (and sometimes to get into trouble doing so), more frequent student-teacher interactions might help vent in a positive way this natural curiosity, especially if teachers use such interactions for academic instruction (show such students where they have made valid generalizations, etc.). Thus, instead of disrupting their own and other's work these students have an acceptable (and reinforced) outlet for their energy in structured classrooms. If properly structured by the teachers, increased teacher-student contact can help maintain on-task behavior and reduce overall class disruption.

Independent students' attitudes were generally enhanced by the treatment program, with the exception of eighth-grade students in individualized settings who had teachers we labeled secure and independent. The former finding is quite possibly an aberration in the data and we have no explanatory theories for this occurrence. The finding, which is consistent across the three grade levels (although some interpolation is necessary because of missing data), seems to make conceptual sense and can be explained in two ways. First, because the active teaching program called for more structure and teacher control, one could predict that independent students who were used to a moderate degree of freedom in individualized teachers' classrooms (at least until the treatment program began several weeks after school had started) would not respond positively to a situation that severely restricted their independence. Second, it is possible that when the teachers altered their instructional techniques to conform to the practices prescribed by the treatment, they disrupted established procedures, including student conduct regulations. Considering independent students' propensity to cause behavioral problems, it is quite likely that as the rules changed these students were the first to be disciplined, which would in turn likely promote negative attitudes. Nevertheless, although all the classrooms in the experimental group had to make instructional adjustments after the school year had begun, with the exception of classrooms characterized as individualistic, most changes resulted in more positive attitudes of students labeled independent. This implies that once an individualized setting is established and the teacher has delegated some decision-making power to students, it is difficult to regain authority without negatively affecting independent students' attitudes.

A second general finding was that a typology of students we labeled as high achieving (recall typologies include more information about students than just achievement) profited most from classrooms characterized as fast-paced, teacher-centered, structured, and demanding. Again, there is consistent support for this conclusion from a number of sources. In their study of sixth-grade mathematics achievement, Whitzel and Winne (1976) concluded that high achievers performed better in traditional classrooms because of the faster pace and increased competition. Similarly, Peterson (1976, 1981), Bennett (1976), and Solomon and Kendall (1976) also reported that students they described as prior high achievers functioned best in highly structured, high-participation, fast-paced, and formal classrooms. However, there seem to be at least two mediating variables that influence these relationships. One is the natural tendency of high achievers to quickly adjust to a variety of learning conditions so that these conditions have a lesser effect on them than on other types of students. Thus it is likely that if teachers can make academic goals reasonably clear and provide appropriate learning opportunities, high-ability students will profit from most instructional programs.

A second factor which mediates classroom characteristics and student achievement is more related to the pace and quality of instruction. For instance, asking teachers to provide more structure and higher participation will probably be *inadequate* for increasing student achievement unless the quality of instruction (e.g., form of student participation) and the pace of instruction are also addressed. Even though a very structured, high-participation, formal lesson format is followed, little additional learning is likely to occur unless content coverage is also adjusted. We observed this phenomenon in the fourth-grade study, in which high-achieving students performed only somewhat better in treatment than in control classrooms. This may be because the active teaching treatment model called for teachers to increase content coverage somewhat and most of the high achievers were likely already comprehending the material (regardless of the instructional mode). Had instructional pace increased more for these students, greater gains would have probably been evidenced.

Janicki and Peterson's (1981) findings that prior high-achieving students did better in a variation of the active teaching model that allowed for more independence and choice of homework, within a small-group setting, also point out the possible importance of adapting the model for high achievers. The superior achievement scores of highs in the small-group variation of the model over the regular, direct instruction variation can be attributed, at least in part, to a difference in content coverage. Because high-ability students in the small-group setting were allowed greater control over their own learning, they may have chosen more difficult and interesting tasks. This may have increased the content they covered beyond that encountered by the direct instruction group, whose assignments were chosen by the teacher and probably directed to average students. Also, the chance to work with each other and to teach other students *actively* may have helped these students. Related research has illustrated that under certain conditions peer instruction in small groups can facilitate achievement (e.g., Webb, 1982; Peterson and Wilkinson, 1982).

We found that affective reactions of high achievers to the treatment program and to the teacher typologies were mixed and did not lead to any meaningful interpretation. In addition, because of the low number of students in some categories, our data do not provide a rigorous test of attitude-related hypotheses. Although some investigations (Janicki and Peterson, 1981) have reported that high-ability students had more positive attitudes toward math in direct instruction than in a small-group variation of the direct instruction model, such results probably are merely suggestive at this point. Because of the importance of maintaining positive student attitudes, however, further study in this area needs to be undertaken.

A third major finding suggests that students characterized as dependent function best with sympathetic teachers in rather teacher-centered classrooms and with teachers who employ techniques typically suggested by the active teaching model. In contrast, dependent students and extremely low achievers do poorly academically with teachers who are subject oriented and emphasize academic accomplishment over social concerns. Dependent students probably need more teacher attention, including more specific instructions and feedback than they are likely to receive in achievement-oriented classrooms. This finding, although somewhat speculative, seems to give added support to those who advocate matching students and teachers on the basis of similar characteristics. It seems natural that a symbiotic relationship would develop between teachers who lack security in teaching math, for instance, and students who need teacher support. Although this prediction seems plausible, it is not as well documented as is the appropriateness of the active teaching model for dependent students. In reports which suggest appropriate classroom environments for students with varying characteristics, the most frequent recommendations are that dependent students need structured, teacher-centered classrooms in which there is opportunity for frequent teacher-student interaction and where rules and objectives are clearly delineated (Solomon and Kendall, 1976; Bennett, 1976).

When we examined the attitude residuals from dependent students across grade levels we discovered some interesting discrepancies. For example, in the fourth grade, the direct-instruction treatment generally had a positive effect on dependent students; however, in the eighth grade this treatment had just the opposite effect. In addition, the teacher types that produced the largest positive attitude change in the fourth grade (unsure and secure), produced negative attitudes in the eighth-grade and to a lesser degree in the sixth-grade sample. Assuming that our measures were reliable and our methodology correct, which we believe to be the case, then a strong context effect was obviously present. From our observations of the trends in teachers' and students' responses to the student aptitude and teaching style inventories, such a context effect was somewhat predictable.

The active teaching treatment program can benefit dependent students by providing a more stable, predictable classroom and affords the opportunity for more teacher contact. Yet it can also reduce individual student-teacher personal contact, since the teacher is spending more time with the class as a whole and less time with individual students. Considering that dependent students seek teacher support, are most likely to ask questions and initiate other teacher contacts, and probably occupy most of the teacher's time, any reduction in private or public opportunities for such interaction would likely be stressful for these students and might impair their performance, at least temporarily. Because lower-grade classes are likely to be less structured and formal, the treatment program is beneficial in that it adds predictability and stability, and indeed one could argue that although the active teaching model reduces opportunities for personal contact between the teacher and a student, total substantive contact is increased. In contrast, upper-level math classes are reasonably structured, and thus an active teaching program probably reduces even further the number of private, and to some extent public (e.g., more lecturing and fewer unsolicited questions), teacher-pupil interactions. Thus, while more structure might be beneficial in the lower grades, any increase above a reasonable amount probably results in fewer teacher-pupil contacts. This reduction in interaction more than offsets the desirable effects of increased structure. Consistent with this explanation, we found that the most positive and negative attitudes of dependent students were associated with the best program implementers at the fourth- and eighth-grade levels respectively. That is, students of teachers who implemented most aspects of the active teaching model had positive attitude changes at the fourth grade but negative changes at the eighth grade. In terms of dependent students, then, there seems to be a nonlinear relationship between the amount of structure which characterizes an active instruction model and their attitudes. Teacher-centered behavior positively affects student attitudes only to a point, then it produces negative attitudes. For teachers and researchers who choose to employ the active instruction model in upper-grade-level classes, it seems important to first assess the amount of structure already present in classrooms and then to make adjustments in the model. Merely providing a more defined, structured, and teacher-centered environment than that already present is inadvisable if attitudes of dependent students are to be considered. As stated previously, teachers have to adjust the program to the particulars of their teaching situation.

In many ways our findings related to the achievement of low achievers and withdrawn students parallel those for students we labeled dependent and are among the most significant results of our work. In general, lows profited most from warm environments where teachers provided task-oriented but nonevaluative support. They also were aided by active teaching which emphasized review (thereby providing increased structure, and probably reassuring low-achieving, withdrawn students to some extent). (See Good, Ebmeier and Beckerman, 1978, for a discussion of effective teaching strategies in high- and low-SES classrooms).

Other researchers' descriptions of advantageous environments for low achievers are similar to ours, with the exception that they emphasize the need for greater individualization (Whitzel and Winne, 1976; Bennett, 1976). Although this emphasis may seem contradictory to our active teaching model, the two probably are interrelated through another factor we would term appropriateness of content. That is, we believe that the direct instruction model is effective for low achievers, but only if teachers select *content*, examples, and instructional techniques appropriate for those students. Just as prior high achievers do not benefit from coverage of previously learned or easily understood material, if low achievers encounter problems because of inadequate prerequisite skills then any instructional method, no matter how efficiently or skillfully executed, will be of little benefit. Thus, if low achievers clearly will not profit from instruction directed at "typical" students (a case more common at the upper-grade levels), then teachers are probably well advised to establish more self-paced programs for individual pupils or to institute some form of grouping. If, however, teachers can modify active teaching techniques to include appropriate objectives and practice exercises for low achievers (and at the same time direct most instruction to the average students), then the active teaching model would be appropriate, and grouping or special lessons would not be necessary. Of course, the advantage of teaching the class as a whole is that it is more efficient and teachers can engage in more active teaching and do not spend time in noninstructional activities typical of individualized programs (e.g., organizing learning stations, maintaining accountability systems, preparing a multitude of lessons).

Low achievers' affective responses to active teaching were generally positive at the fourth-grade level, but somewhat mixed at the upper grades. The lower-grade gains likely are related to these students' improved academic achievement. That is, because key attributes of the active teaching program had a positive effect on low achievers, they found themselves able to complete more assignments, answer more questions correctly, etc., and thus had better general attitudes. Comparable data at the secondary level again illustrate the possible negative consequences of calling for excessive structure with low-achieving students. Structure apparently assists low achievers to a degree, as evidenced by the fourth-grade attitude and achievement scores; however, in the upper levels lows are typically several grade levels behind, so that more deliberate teacher control does not help much without a concomitant change in course content. Regardless of the appropriateness of the instructional technique employed, if the content selected is inappropriate, little learning will occur and negative attitudes will result. Structure clearly had a nonlinear relationship with achievement over the grade levels for prior low achievers, at least as classrooms were structured in our study.

A final major finding, and one of great potential significance for those charged with implementing educational programs, involved the interaction between teacher types and the treatment program. Our investigations and those of Janicki and Peterson (1981) suggest that teachers who implement the model get good results, yet some teacher types choose to use more facets of the model than other teacher types. This phenomenon is probably a result of the different ways teachers perceive their roles, their previous experiences, and the expectations of their particular schools. For instance, people will more likely adapt and internalize ideas that are consonant with their existing beliefs. Furthermore, many teachers have no firm ideas regarding optimal ways of teaching mathematics. Thus, one would predict that teachers who already believe in an active teaching model or teachers who are unsure using their present instructional strategies would be most likely to implement the experimental treatment program. Indeed, our results in the fourth- and eighth-grade studies, where implementation data are available, support these hypotheses. For example, Type 3 teachers (educated/secure) in the fourth grade and Type 3 eighth-grade teachers (secure/concrete) who indicated that they taught in a more direct, concrete manner did indeed implement more of the treatment program than teachers in the fourth and eighth grades who preferred to teach using an individualized model. Similarly, Type 2 fourth-grade teachers (experienced/unsure) and Teacher Type 4 (unsure) in the eighth grade who indicated they were insecure teaching math in their present manner consistently employed more of the treatment behaviors than did teachers who were secure, relatively inexperienced, and taught in a more abstract fashion. These latter teachers apparently were still utilizing the teaching strategies they learned in their college mathematics methods courses, were comfortable using these techniques, and saw little need to change.

FUTURE RESEARCH

The value of the program for particular types of students is far from clear at this time and more research is needed if a better understanding of the relationship between various parts of the program and different types of teachers and learners is to be achieved. Many questions need to be pursued in future research. For example, we have assumed that teachers who implement the program more fully are either looking for new teaching methods or are already teaching in a way consistent with the program. However, it is important to assess teachers' behavior prior to their being introduced to the program.

Researchers might also vary one dimension of the program while holding others constant, in order to assess the effects of the modified program on students generally or for specific types of students. One could conduct such an experiment by having several teachers use the basic Missouri effectiveness program, but some teachers would check and discuss homework in class (as recommended by the program) while others would not. As Slavin (1982) notes, we might find that specific instructional interventions (components) might be differentially effective with different grade levels, content (fractions vs. long division), or types of teachers and/or students. The variations possible are quite numerous, because any lesson part could be varied or eliminated. However, such research would be helpful in explaining *why* the program works in particular contexts.

Ultimately, if we are to better understand why the program works we will need to have alternative treatment procedures that are fully conceptualized and supported by research. For example, rather than just comparing two groups of teachers who use most of the Missouri program but vary a single dimension (mental computation), we need to do research on alternative mental computation teaching models to be certain that our model is the most effective. Often researchers in the ATI tradition spend much time in identifying student and teacher characteristics, but little effort is spent on developing programs related to those needs (Good and Stipek, in press).

Hence, it is relatively easy to determine if a part of a program is necessary (eliminate that part and verify what happens), but to make such a determination it is first necessary to generate effective alternative programs. To say "give students more choice" or "encourage students to work in small groups" is gratuitous unless we can specify the dimensions from which students can choose or the process that should be present when students work together. There is currently some information about effective large-group instructional techniques in fourth-grade classrooms. When research yields information about effective small-group teaching in fourth-grade classrooms, questions about the effects of alternative treatments on different types of students and teachers can be raised. Without alternative instructional procedures, however, it may be impossible to study interaction effects.

Obviously, teachers must adapt our program to their circumstances. The teacher in an inner-city high school may be confronted with many students who prefer individual seatwork assignments and who actively resist public interactions in the classroom (e.g., Metz, 1978). Such a teacher will have students who have become intellectually passive and who do not listen to teachers (e.g., Good, 1981). Thus, initial assignments in such a setting would be directed at getting students to listen. Our advice to teachers who are interested in using the program is to follow the program for a year as it appears in Chapter 3 and to keep careful records of pace, activities, and student work. Teachers could then determine how subsequent changes in the program affect their students' attitudes and work.

CONCLUSION

We found that most teachers who implemented the program obtained positive gains from students; however, it is not clear why some teachers implemented the program better than others. It may be that some teachers cannot use the program (insufficient mathematical knowledge, no confidence when working with a large group of students); others may be more effective using other instructional styles. Only more research that involves the development of more advanced training procedures as well as alternative instructional procedures will clarify this issue.

The treatment has been tested only in its entirety, and certain aspects of the program may be unnecessary. Future development work should examine the effects of adding or deleting program parts on student achievement. We have encouraged teachers who want to use the program to first implement the program in its recommended form and to then adjust it to meet needs identified in their classes (and to assess the effects of the adjustment).

Each student is unique and needs instruction appropriate to his or her level. However, the difficulty of examining the effects of instruction on every student is so overwhelming that we have used student and teacher typologies as a means of considering instructional techniques (or programs) as they relate to classroom context. To some extent these typologies are artificial because they are composed of an average of many traits. Furthermore, preferences or quite different variables can be used to form types and to assess instructional consequences.

The research we have done on the interactive effects of the treatment with different types of students and teachers is quite exploratory. The results we have obtained do not indicate how the program should or could be adopted. However, the preliminary data presented in this chapter do illustrate some contradictions (e.g., older students' increasing interest in choice and their teachers' decreasing interest in providing choice) that may encourage teachers to consider how and why the program should be modified in certain contexts.

9

The Role of Development in Mathematics Teaching

The development portion of a mathematics lesson plays a key role in the active teaching model we have developed and researched over the past decade. This research has shown that development is a complex process that interacts not only with other phases of a lesson, but also with many context variables. For example, the amount of time teachers devote to development seems to be closely associated with student achievement at the elementary school level, while quality of development seems to be a more pertinent issue at the secondary school level. It now appears that this is because elementary teachers generally allocate a larger portion of their mathematics time to practice activities than do secondary mathematics teachers. Because the average elementary school teacher spends little time on development it appears that requests to increase the amount of time on development (however minimally) are associated with improved teaching. In secondary classes, many teachers spend somewhat more time on activities that they believe to be development. Increases in time are less likely to influence achievement at this level unless more emphasis is placed upon the quality of development.

Our past research has led us to regularly reconsider this concept and how it might be best defined, measured, and used in mathematics classrooms. This chapter presents evidence to support the position that development is an important part of a mathematics lesson; a summary of our past efforts to characterize and measure development; and our most recent thinking about key aspects of development and the most productive direction for research in this area.

HISTORICAL OVERVIEW

Development is a part of most mathematics lessons in which the teacher actively interacts with pupils. In very general terms, development can be thought of as that part of the lesson devoted to the meaningful acquisition of mathematical ideas, in contrast to other parts of the lesson such as practice or review.

Many teachers do not generally devote much time to development. For example, Dubriel (1977) found that a group of first-year algebra teachers spent an average of only 25 percent of their class time on development, although there was considerable variance among teachers. Personal observation in classrooms and discussion with other classroom observers also suggest that not only does the quantity vary, but the quality of development differs considerably from teacher to teacher. It is interesting to note, however, that when educators are asked to do so, few can identify specific criteria used for evaluating development.

Insufficient attention has been given to the quality of development in our work and in educational research generally. This may be related to the absence of an operational definition of the term in the literature. Still, several research studies (e.g., in mathematics classrooms) have examined the relationship between amount of development in mathematics instruction and student achievement. The definitions of development utilized in these studies are quite brief and lack detail. In fact, they are very inadequate for addressing issues of quality or developing training programs. However, these studies find a consistent and strong positive effect on pupil achievement associated with increasing the amount of time devoted to developmental work within mathematics class time.

Shipp and Deer Study

Shipp and Deer (1960) investigated whether varying the percent of class time spent on development and on practice work affected mathematics achievement. They defined development activities as primarily group activities intended to increase understanding and general usefulness of the ideas studied. These activities included teacher explanations, demonstrations, and class discussion; visual and manipulative materials; group reading; drawing; construction work; and project work. Practice work included individual pencil-paper work on computation or verbal problems, or other activities or questions taken from the textbook. These activities were to be completed by pupils individually.

In fourth-, fifth-, and sixth-grade classrooms ($N = 12$) where teachers were carefully matched, Shipp and Deer experimentally compared four proportions of development and practice time: 75 percent development/

25 percent practice; 60 percent development/40 percent practice; 40 percent development/60 percent practice; and 25 percent development/75 percent practice. The treatment period lasted 12 weeks and daily mathematics class time was 45 minutes. Teachers were not required to include the requested amount of development each day, but rather to adjust time allocations cumulatively.

The results of this study showed that groups spending more time on development than on practice had higher total achievement scores in grades four and five and produced higher scores on "understanding" and "using" subtests in grades four and six. The treatments did not produce differences in problem-solving scores. Further, the authors found that amount of developmental work did not interact with three levels of pupil ability. That is, more time on development increased the achievement of high-, middle-, and low-achieving students.

Shipp and Deer concluded that in this study higher mathematics achievement was associated with increased time spent on development, and that spending more than 50 percent of class time on development seems appropriate for pupils at all ability levels.

Shuster and Pigge Study

Shuster and Pigge (1965) attempted to determine how much class time should be devoted to "developmental-meaningful" activity and how much to "drill or practice." They defined these approaches in this way:

> The term "developmental-meaningful approach" was used to refer to the method of teaching which leads students to conclusions by means of thought processes utilizing meaningful and/or socially significant arithmetic experiences. The teaching activities that comprised the portion of the time assigned to the developmental-meaningful approach included: (1) explanations, demonstrations, and other developmental work presented by the teacher, textbook, or pupils; (2) discussions led by the teacher; (3) use of audio-visual materials; and (4) use of manipulative materials.
>
> The term "drill approach" was used to signify the method of teaching in which systematic repetitive practice with examples and problem-solving situations is presented authoritatively. The teaching activities that comprised the portion of the time assigned to the drill activities included: (1) oral questioning by the teacher or other activities, such as flash cards, to elicit automatic immediate responses from the pupils; (2) pupils working with examples from drill booklets specially prepared for the study; and (3) pupils' work with problems and examples presented by the textbook, various workbooks, and teachers. (p. 25)

In this study three ratios of development to practice were examined: 75 percent development/25 percent practice; 50 percent development/50 percent practice; and 25 percent development/75 percent practice. Eighteen randomly selected fifth-grade teachers participated, with six teachers assigned to each treatment condition. The treatment lasted 22 days and time allocations were adjusted on a cumulative basis. Both an immediate posttest and a retention test (six weeks later) were given.

The results of the study showed no differences among treatments on the posttest results. On the retention test the 75 percent and 50 percent development groups performed significantly better than the 25 percent development group on both the total test and the computation section. There were no differences in scores on verbal problem solving. Shuster and Pigge concluded that spending more class time on developmental activities lead to increased retention.

Zahn Study

Zahn (1966) experimentally examined the use of class time in eighth-grade mathematics classes. Four treatment groups were randomly formed from a sample of 120 students and used to determine the effect of varying the percent of class time devoted to developmental activities and to practice work. Developmental activities were defined as those intended to increase understanding and included teacher and class demonstrations, explanations, group reading, and discussions. Practice work included individual pupil pencil and paper work on assigned exercises and would be considered drill.

Forty-five-minute class periods were divided into development and practice portions for each treatment group as follows: 67 percent development/33 percent practice (Treatment A), 56 percent development/44 percent practice (Treatment B), 44 percent development/56 percent practice (Treatment C), and 33 percent development/67 percent practice (Treatment D). The treatment period was 18 weeks in length and at the end of the instructional period a three-part posttest was given.

Analyses of the data showed that in several cases Treatment A and Treatment B groups scored significantly better than Treatment C and D groups on the subtests of the posttest. In no case did groups C and D score significantly higher than groups A and B. The author concluded that a teacher should spend 50 percent or more of a class period on developmental activities.

Zahn also examined the relative effectiveness of the treatments on three ability levels of pupils within the treatment groups. In each of the four groups pupils were classified as upper-, middle-, or lower-ability according to IQ scores. In general, Zahn found, as did Shipp and Deer, that pupils of various ability levels were not differentially affected by the variations in class time. The one exception was that high-ability pupils in Treatment A (67 percent development) gained significantly more than high-ability students in the other three treatment groups on total arithmetic achievement.

Dubriel Study

Dubriel (1977) studied utilization of class time in 15 first-year algebra classes. Five sections were taught using a 70 percent/30 percent ratio of development to practice, five sections used as 30 percent/70 percent ratio, and five sections acted as controls and were post hoc determined to average about 25 percent development and 75 percent practice. The study involved a six-week treatment period, during which mathematics content was carefully controlled through the use of daily lesson plans provided by the researchers. Classroom observation was used to ascertain conformity to treatment conditions and later analysis of this data showed that fidelity to treatment was excellent. A posttest composed of subtests measuring knowledge, skills, and problem solving (word problems) was administered to all students at the end of the treatment period. To measure retention all students were retested seven weeks after the posttesting with a parallel form of that achievement test.

When the achievement data were analyzed there were no significant differences among the three groups on the knowledge and skills subtests. On the problem-solving test, however, the 70 percent development/30 percent practice group and the 30/70 group both scored significantly better than did the 25/70 group. Although the 70/30 group mean was higher than the 30/70 group mean, the difference was not significant. This pattern was also present in the problem-solving retention data, where the 70/30 group mean was significantly higher than the 30/70 group mean, which in turn was significantly higher than the 25/75 group mean.

Dubriel noted that these problem-solving results favoring more time on development to improve problem-solving performance were consistent with Zahn's findings in eighth-grade mathematics classrooms. He also stated that the achievement gains of the younger students studied by Shipp and Deer and Pigge and Shuster were in cognitive areas other than problem solving. Dubriel concluded that more teaching time should be spent on developmental activities in first-year algebra classes.

DEVELOPMENT IN THE MISSOURI MATHEMATICS PROJECT

We initially attempted to define development when we conducted a naturalistic study designed to identify instructional behaviors of fourth-grade teachers who consistently produced large pupil gains in mathematics on standardized achievement tests (Good and Grouws, 1975a). The same paradigm was used to examine teaching behaviors of relatively ineffective teachers. Based on the previously described experimental studies and our earlier classroom research, development was chosen as an important variable and relevant data were collected by classroom observers.

As noted above, the definitions of development used in previous experimental studies were not very detailed and, consequently, were of only limited help to us in defining the term. Due to the size of the task and the limited amount of time available to us, we concentrated on defining development so that we could quantitatively measure it in the classroom.

Videotapes of classroom mathematics teaching and lengthy discussions of development based on the teaching experience of the research team were used to help formulate a definition of development. For purposes of classroom observation we summarized development in this way:

> The *development* portion of a mathematics period is that part of a lesson devoted to increasing comprehension of skills, concepts, and other facets of the mathematics curriculum. For example, in the area of skill development, instruction focusing on *why* an algorithm works, *how* certain skills are interrelated, what properties are characteristic of a given skill, and means of estimating correct answers should be considered part of developmental work. In the area of concept development, developmental activities would include initial instruction designed to help children distinguish the given concept from other concepts. Also included would be the associating of a label with a given concept. Attempts to extend ideas and facilitate transfer of ideas are a part of developmental work.

In coding the amount of time allocated to development work observers were instructed not to include the following as developmental: transition time, time going over homework, review time, drill, and homework practice. Our estimates of amounts of development time were therefore conservative because development is often a part of many of the activities that were excluded by our criteria. For example, a great deal of development can occur when homework is checked, if teachers provide explanations, examples, and so on, in addition to correct answers. In this same way, development can be an integral part of review activities. An interest in a measure of "pure" development and the desire for a highly reliable coding scheme led to the aforementioned criteria, which tend to underestimate the amount of time spent on development.

The results of this study showed that effective teachers spent more time on development than relatively less effective teachers. An average of about 16 percent of class time was allotted to development by the effective teachers, and a bit less than 13 percent by the relatively less effective teachers. Note that the amount of class time devoted to development (using our conservative definition) was very low, and Dubriel's estimate of 25 percent may be a more accurate figure. In addition to spending slightly more time on development, effective teachers also used development time in a more active and appropriate way. They generally provided additional developmental work prior to practice and seatwork activities. In some classes this procedure provided a smooth transition from the teacher's major presentation of material to practice work. In other situations the additional time teachers spent explaining may have increased students' success on the practice work, which may have accounted for the low number of behavioral problems. The additional development may also have increased pupils' time on task. Because development can potentially affect many phases of a lesson, we can only hypothesize concerning why more time spent on development is associated with effective mathematics teaching. Additional research is needed before these hypotheses can be substantiated.

Development was a prominent part of the active teaching model that we experimentally tested in 40 fourth-grade classrooms in 1977. From previous work we knew that a large segment of development work within a mathematics lesson was necessary, and that this segment should relate to other parts of the lesson. As a result, we developed a framework, or set of guidelines, for organizing and teaching mathematics (see Chapter 3).

We were unsure of how many new instructional techniques teachers could reasonably be asked to utilize in a very short time and with minimal training. Even willing persons can only be expected to concentrate on a limited number of things when actively performing a multi-faceted task. This is especially true of teachers, who must constantly react to the complex, changing conditions in classrooms.

Two decisions were made with regard to development. First, treatment teachers were asked to spend about one-half of the mathematics class time (about 20 minutes) on developmental work. Second, teachers were asked to increase student participation in development just prior to the practice portion of the lesson. However, teachers were not specifically asked to increase the *quality* of development and no specific guidelines or suggestions for improving development were given. At that time it was not entirely clear to us what the components of development were, let alone how they could be improved.

As was noted in Chapter 4, the treatment behaviors were well implemented in general, and treatment teachers' profiles were significantly more similar to the requested behaviors than were the profiles of control teachers. Yet many treatment teachers did not reach the criterion of 20 minutes of development daily (using our conservative definition of development). An examination of all the behaviors requested of treatment teachers clearly indicated that development was probably the most difficult portion of the program for teachers to implement, from both a conceptual and practical standpoint. Considering the higher levels of implementation for the other requested behaviors, it seems likely that treatment teachers tried to implement development (e.g., needed more specific advice) and could not. It is also possible that some teachers attempted to maximize the number of program behaviors they employed, and that when this became too demanding, they chose to ignore development because it was the most complex behavior.

The achievement gains for the treatment-group students were quite dramatic, and we cannot help but wonder whether the gains would have been even larger if fidelity to the development dimension had been better. In order to study this latter possibility, teacher training time would probably have to be increased, because we sensed that some teachers were not cognizant of meaningful ways to spend more time developing simple mathematical ideas.

Development was also a part of our experimental research in junior high mathematics classrooms (see Chapter 6 for details of this research). Treatment teachers were again asked to devote a large part of their instructional time to development, but an important change from the fourth-grade study was the addition of a daily, ten-minute verbal problem-solving component.

We intended this ten-minute segment to be spent on development of problem solving, with the teacher actively interacting with the class. We hoped that these instructional recommendations would lead to increased time spent on development. However, anecdotal records from observers indicated that whether or not development was a part of the problem solving, as well as the quality of the development, varied from teacher to teacher.

Still, we found that development was the most difficult aspect of the program for teachers to implement, just as it had been in the fourth-grade study. Treatment teachers varied considerably in amount of development time; some did not spend the recommended time on general development, and others spent considerably less than the daily ten minutes on problem solving. This result was unexpected; neither our consultants nor the partnership teachers (nor we!) thought that teachers would have difficulty utilizing the recommended amount of time for development. Although it is commonly believed that most secondary mathematics teachers spend half of each instructional period on development, our data suggest that when development time is conservatively measured, considerably less time is used.

Although treatment teachers did not implement the development requirement as well as expected, their mean total development time (general development plus verbal problem-solving time) was considerably more than the total development time for control teachers. This difference was due mainly to the extra time treatment teachers spent on verbal problem solving. This may partially explain why the treatment-student group performed significantly better than the control group on the problem-solving measure and slightly better on the general mathematics achievement measure (see Chapter 6). If our ability to define and clarify the concept of development improves, better implementation of development and increased student achievement may follow.

From our research we have concluded that development is an important aspect of instruction because it appears to relate positively to student learning, although many elementary and secondary teachers have difficulty increasing the time they spend on development. There are many possible explanations for this difficulty, including insufficient teacher *motivation* (active teaching requires preparation and is physically demanding); *new instructional procedures* (teachers are asked to make too many changes too quickly); or researchers' inability to accurately define the *content* of development (better qualitative descriptions are needed). An operational definition of development may or may not be needed to prepare improved training programs; however, it is definitely requisite for useful research in this area and for the design of more adequate theoretical models. Our initial efforts to detail the qualitative aspects of development are described in the next section of this chapter.

DEVELOPMENT AND TEACHING

Mathematics teaching is enjoyable in part because it involves a variety of tasks. Teachers often manage, diagnose, prescribe, evaluate, organize, record, produce, encourage, edit, develop, and reteach when conducting a mathematics lessons. How teachers approach each of these tasks provides a useful framework for describing what takes place in classrooms and a means of studying mathematics teaching that has great potential for improving the quality of instruction. While there is some agreement among educators about what behaviors each of these roles requires, there is no widely accepted definition of any of them. In this section we will expand our original definition of development and discuss the importance of this concept to teaching and research on teaching.

It is often easier to give meaning to a term by stating what it is *not* rather than by describing its relevant attributes. This is true of definitions of development, and perhaps explains why previous descriptions have included phrases like "not the practice portion of the lesson," and "involves things other than individual work on drill activities." Much is lost in such an approach, however, because when one wants to discuss the quality of development in instruction there are no apparent components of the activity to examine, refine, or analyze. Therefore we first characterize development by using descriptions and examples, and then we will try to clarify the concept by giving negative instances and contrasting it to related terms and ideas. Although we cannot offer a comprehensive definition of development, we hope to provide a more detailed conceptualization of development than has been available previously.

DEFINITION OF DEVELOPMENT

Development in mathematics instruction is the process whereby a teacher facilitates the *meaningful acquisition* of an idea by a learner. Meaningful acquisition means that an idea is related in a logical manner to a learner's previously acquired ideas in ways that are independent of a particular wording or special symbol system. Thus, development can be conceptualized as a collection of acts controlled by the teacher that promote a particular kind of learning. The identification of these elements is important.

Five components that we believe to be important parts of development are discussed in the next section. It should be noted that not all components of development will necessarily occur in any particular lesson. Likewise, the existence of one component does not guarantee that development is taking place. Development is a very complex phenomenon and the particular elements that constitute successful development may vary from one teaching context to the next. It is also important to realize that effective development may not be composed of the same combination of behaviors, even in similar classroom environments.

COMPONENTS OF DEVELOPMENT

Attending to Prerequisites

In mathematics, an understanding of most new concepts and processes is based upon previously learned ideas and skills, although the prerequisite ideas/skills vary depending on what is to be learned. For example, it would be very unlikely for a pupil to acquire meaningfully the notion of a prime number without some knowledge of factors (divisors). Similarly, proficiency in using the traditional long division algorithm is not possible without subtraction skills. It is therefore important for a teacher to take account of prerequisites in conducting development.

Our work has suggested that attending to prerequisites in the development phase of the active-teaching model is necessary. Prerequisite concepts can be taught prior to the introduction of a new concept, or work can begin immediately on a new idea and prerequisite ideas can be given attention as they are needed. An illustration of a direct way of attending to prerequisite concepts in a lesson on fractions follows:

Teacher: Adding fractions with unlike denominators won't cause much difficulty if we remember a few basic ideas. Let's review some of them. Can you think of another name for 1/4?
Student: Two-eighths.
Teacher: Okay, other names?
Student: Three-twelfths, four-sixteenths.
Teacher: That's right, what about 20/80?
Student: Yes, because if you divide the top and bottom by 20, you get one-fourth.
Teacher: What do we call fractions that name the same number?
Student: Equivalent fractions.
Teacher: Yes, can someone show . . .

Attending to Relationships

Mathematics is composed of a large body of logically consistent, closely related ideas. To teach it as a collection of isolated facts to be memorized (as is often done) is a disservice to the learner and to the discipline. Relationships among mathematical ideas are numerous and the development portion of the lesson provides the teacher with an opportunity to emphasize the meaning and interpretation of mathematical ideas. Some mathematical relationships are simple, such as the doing-undoing relationship between addition and subtraction, or the relationship between angle measure and side length in a triangle. Other relationships are more complex: Is the product of two numbers greater than their sum? Is there always a "next" fraction as there is with whole numbers? As part of the development the teacher brings into focus the interrelatedness of mathematical ideas and thus stimulates interest, increases retention, and enhances the likelihood of transfer of an idea to new situations.

Consider the teaching of fractions again. When properties of addition of fractions are related to previously learned properties of addition of whole numbers, connections become apparent and learning is made easier. The addition properties that are satisfied under both number systems, for example the commutative property $(a + b = b + a)$ and the associative property $(a + b) + c = a + (b + c)$, begin to provide a unifying strand for the mathematics content. The possibility of conceptualizing the whole numbers as a part of the set of fractional numbers is suggested, and other important relationships emerge as well. Emphasizing relationships increases retention and also allows students to easily reacquire forgotten skills or concepts. The following excerpt depicts a teacher eliciting information about relationships from students as part of a class discussion of fractions.

Teacher: That's very true, can anyone else think of a relationship between fractions and whole numbers?

Student: The order in which you add two fractions doesn't change the answer, just like with whole numbers.

Teacher: Yes, any others?

Student: Whole numbers are greater than fractions.

Teacher: Can you give an example?

Student: Sure, two is more than one-half, three is more than one-half; all of them are more than one-half.

Teacher: Is 5/4 a fraction?

Student: Yes.

Teacher: Can you show me where it would be on this number line?

Teacher: That's correct. Now is 5/4 greater than one?

Student: I see now.

Teacher: Any other relationships between fractions and whole numbers?

Student: You can write any whole number as a fraction by writing the whole number over one.

Teacher: Yes, that's right. What about this... is there a fraction between any two whole numbers? Is there a whole number between any two fractions?

Student: No.

Teacher: "No" to which question, or did you mean "no" to both?

Student: Well, there is a fraction between any two whole numbers but not the other way around.

Teacher: Can someone give me an example?

Attending to Representation

Mathematical formulas, theorems, processes, and indeed most of mathematics, are a step removed from the physical world. This level of generality of mathematics contributes to its great power, because any single mathematical idea or process can accurately represent a great variety of real-world phenomena. For example, a fraction like 1/4 can convey important information on an advertisement, recipe card, gauge, sales brochure, time sheet, and so on. The abstract nature of mathematics which enables it to serve as a model for widely diverse physical phenomena may also create substantial impediments to learning. This is especially true if a teacher presents a mathematical concept in its abstract form without giving examples of its concrete representations. A teacher might use concrete representations to aid students' understanding of fractions in the following manner.

Teacher: Can you think of some other situations where we use one-fourth or how we have represented one-fourth?

Student: We use one-fourth in telling time, like a quarter past one.

Teacher: Yes, any others?

Student: Measurements use one-fourth. Like you can have one-fourth inch of rain.

Teacher: Right. About how many of these books are paperbacks? Would that be one-fourth?

Student: Yes.

Student: The desks in my row would be one-fourth of the desks in the room.

Teacher: Is that correct, class? How many . . . .

Attending to Perceptions

As students manipulate materials or observe teacher demonstrations they become aware of the *relevant* aspects of the mathematical concept under study. Equally important, they discern attributes that are present but are *not relevant* to a concept. Through sensory input they may form a mental image of a mathematical idea and relate it to previous learning. The formation of these links can be promoted by the teacher through the use of questions that ask students to relate a new concept to previously learned ideas. Some questioning can be more specific and lead students to discover a particular relation and verbalize it.

The comprehension acquired through perception is important; however, to make certain that students fully understand a mathematical idea, a teacher must also insure that pupils know and can apply the idea *independent* of a particular physical representation. For example, using the Cuisenaire Rods to teach addition of whole numbers may be very appropriate *if* a teacher also develops the idea in connection with other perceptually different materials such as sets of objects, number lines, and balance beams. Only in this way can a pupil be freed of thinking of addition apart from a single representation. This ability to think of a mathematical idea in many different ways is crucial to being able to apply the idea in other learning situations or in everyday life, and helps students to form appropriate abstractness of a concept.

When to introduce a new, perceptually different representation must be carefully decided by the teacher. Too early an introduction causes confusion for learners and a late introduction often causes difficulty because students have become accustomed to associating an idea only with familiar material. The following excerpt illustrates a teacher who is introducing a new representation for fractions.

Teacher: We are going to let part of the class represent part of a number line. Will the front row please come up here and arrange yourselves in a straight line? Bobby, since you're at one end, you will represent zero. And Robin at the other end represents the number one.

Student: What number do I stand for?

Teacher: Well Dave, you're standing exactly in the middle of the line between Bobby and Robin. So what number do you think you represent?

Student: I must be one-half.

Teacher: That's right! Who can tell me which person represents one-fourth?

Student: Sherry does because she is standing half way between Bobby and Dave.

Teacher: That's very good, but what if . . . .

Another example involving a real world representation follows.

Teacher: We have been thinking of fractions as parts of things, but they have other meanings, too. We can use them to compare amounts of things. Here is a bag of M & M's. How many are in the bag?

Student: Forty.

Teacher: Okay, and how many of each color are there?

Student: Ten brown, 5 orange, 3 yellow, 20 tan, and 2 green.

Teacher: That's right. Now, thinking of the number of brown as a fraction of the total in the bag, what fraction is represented by the brown?

Student: 10/40.

Teacher: Correct. What are the other fractions?

Student: 5/40 orange, 3/40 yellow, 20/40 tan, 2/40 green.

Teacher: Does that take care of all the M & M's in the bag?

Student: Yes.

Teacher: Since that's all of them, then adding these fractions together should give us one. Add them up. How many forties are there?

Student: I get 40/40.

Teacher: Yes. And that is one whole bag, so everything seems to be working out. Remember, then, that any time we want to compare a part to a whole it can be stated as a fraction. Whether it is three hits at five times at bat, or 9 of 10 spelling words correct, they can all be expressed as fractions.

Student: Can we eat them now?

Generality of Concepts

Teachers often do not fully explain the general applicability of a mathematical idea or process to students, even within the limits of the students' abilities. For example, teachers can explain that the method used to find the sum of the angles of a quadrilateral works equally well for finding the sum of the angles of any polygon. Similarly, the term vertex applies to all polygons and not just to triangles, as it is sometimes introduced. Clearly, only a few of the applications of a mathematical idea should be introduced in the initial discussion of the idea. Ideally, the concept or skill should be developed in as general a setting as possible rather than artificially confined to a narrow set of conditions. In the following excerpt a teacher explains the applicability of a concept in a lesson on fractions.

Teacher: What do we call the fractions 1/2 and 1/3 in the sentence 1/2 + 1/3 = 5/6?

Student: Addends.

Teacher: That's right, and what about the 5/6?

Student: Sum.

Teacher: Okay. Remember how when we were adding whole numbers we used the same terms? Now, when we added two whole numbers the sum was always larger than either of the addends. Is this true for fractions, too?

Student: Yes.

Teacher: Can you give me an example?

Student: Well in your sentence the 5/6 is greater than the 1/2 and the 1/3.

Teacher: Can someone give me another example?

Student: 1/5 + 2/5 is equal to 3/5 and the 3/5 is greater than the 1/5 and 2/5.

Teacher: You've got it. What happens when we think about multiplication of fractions?

Holistic Viewpoint

Development has been described in terms of its components. This description is not detailed or exhaustive, but rather is an initial attempt to provide a definition for development which will enable researchers to examine the qualitative aspects of development. As we have stated, development is a dynamic process in which different components may interact to promote learning in various contexts.

If development is viewed in a global way, then teachers' attention to the *meaning* of material presented is vitally important, and development must be understood in terms of promoting student *thought*. Future classroom research must study more extensively the *content* that is presented to students rather than instructional time per se.

NEEDED RESEARCH AND FUTURE DIRECTIONS

Our conceptualization of development is a step toward developing a framework for systematic study of this variable. Previous research has demonstrated its importance; it is now time to determine *why* development is associated with increased learning gains. We know that teachers who are performing poorly can probably be guided to improve their teaching by increasing their development time. However, there are few data which suggest to teachers how components of development (such as perceptions, relationships, etc.) interact, or when and to what degree these components should be included in development (Grouws, 1980). In many ways the qualitative investigation of this important concept is just beginning. However, there are several research topics and/or approaches that are likely to increase our knowledge.

Process/product studies which focus on the components of development and how they interact would be useful. In such studies researchers would need to control for the content of lessons. That is, the behaviors which produce effective development may vary from a lesson introducing a new concept to a lesson oriented toward improving verbal problem solving. Care must also be taken in this type of research to insure that global aspects of instruction and behavior are not obscured by data about individual behaviors. Our previous statements about the value of considering *clusters* of related behaviors (see Chapter 2) apply to this situation. Another issue related to this type of research is whether the behaviors associated with effective mathematics teaching tend to be constant across effective teachers, and stable for effective teachers across lessons of the same type.

These latter questions can be addressed very well through intensive observation of, and interviews with, small samples of teachers. Progress in the study of development will come much more quickly if these questions are answered affirmatively. Case studies of this type must consider interactions between lesson parts (homework, seatwork, development) as well as between components of development. As was pointed out before, it is quite likely that the quality and quantity of developmental work influence the length and nature of the other parts of a lesson.

Future study of development will probably concentrate primarily on lessons which introduce new skills or concepts. However, more research is needed on the role of development in lessons related to review of previously taught content, lessons designed to achieve higher order cognitive objectives, and other types of lessons.

Alternate, and perhaps radically different, conceptualizations of development are also needed. It seems obvious that initial conceptualizations will be limited and thus scholarly research on a number of models is desirable. Multiple models will lead to synthesis of results and ideas, and eventually to second-order models that will be more valid, descriptive, and have more implications for classroom teaching than current models.

It will be important to examine and relate findings from research on parallel issues in other disciplines to research on development. The notion of promoting comprehension in reading, for example, is one pertinent area that might have implications for development in mathematics. Development research should also be related to other research on teaching, such as the role of teacher decision making in instruction.

The illustrations of teacher–student interactions that appear above were written to demonstrate individual criteria that can be used to think about development in mathematics and are not intended to illustrate development *per se*. The teacher–student dialogues reflect teacher questioning of individual students (only one aspect of development) and to prevent misunderstanding we want to emphasize that much of development involves teachers' actively developing concepts and explaining to the entire class (see Chapter 3).

10

Conclusions and New Directions

A REVIEW OF OUR RESEARCH GOAL AND FINDINGS

The research presented in this book was based on a program of research supported by three grants from the National Institute of Education. Because of the failure of both educational research and general intervention strategies (not based upon research) to generate meaningful understandings of classroom practice, we decided to observe teachers who were making a difference in student achievement (students' mean residual achievement gain) in a particular context (fourth-grade mathematics). We felt that meaningful variation in teaching behavior did occur and we wanted to test this notion, as well as our general belief that individual teachers make a difference in student learning. Our original intention was not to build a comprehensive mathematics program, but to test the hypothesis that teachers affect student learning.

We chose a standardized achievement test as an operational definition of teacher effectiveness. Although this is not a complete definition of teacher effectiveness (or even an adequate definition), we do feel that it is one aspect of teaching which is important. Standardized achievement scores can be a practical criterion if one understands their limitations and does not overgeneralize findings based upon them. Subsequently, in our program of research we investigated the relationship between estimates of students' learning based upon a standardized test and a content test designed to measure the content teachers presented.

The initial study provided evidence that stable and relatively high- and low-effective teachers could be identified, although many teachers fluctuated from year to year in their "effectiveness" (as measured and estimated by the mean residual gain of their students on a standardized achievement test). From behavioral observation of high and low teachers, it was possible to identify patterns of teaching that differentiated these two groups of teachers. These findings were supported by research elsewhere in field settings and by previous experimental research in mathematics education. These findings, as well as observers' comments about instructional variables in the naturalistic study, were ultimately integrated into a program for training and research purposes.

In our first experimental study, we found that fourth-grade teachers were able to implement the program after minimal training (some trouble was experienced in the development portion of the lesson) and that the degree of implementation was associated with positive student achievement gains. Because the differences in test scores between treatment and control classrooms were large, and because the training time was minimal (six hours—hence, relatively cost effective), the results suggest that the program is a reasonable and practical method of mathematics instruction. Also, the results (both on the standardized test and the test designed to match the content teachers actually presented) clearly show an important teacher effect in these inner-city school classrooms, and suggest that successful educational interventions are possible. However, it was found that the treatment program was better for some combinations of students and teachers than for others.

It is the case that the basic instructional program developed in this project has been well received not only by teachers in our project but elsewhere as well (e.g., Keziah, 1980; Andros and Freeman, 1980). Furthermore, our basic findings have been replicated by others (e.g., Hulleman, 1981; Larkin, personal communication). Such findings suggest that the effects of the program are generalizable and important.

In the second experimental study we developed a problem-solving strategy designed to improve students' ability to work verbal problems which appeared in elementary textbooks. Although the project staff felt that this definition of problem solving was a limited one, we did accept the fact that this was the problem-solving curriculum for many students and teachers. We thus built a program (see Appendix B) that was designed to affect this type of mathematics performance.

In developing the program, we found virtually nothing in the literature to describe teachers' views of problem solving and what they do when they teach such content. It is important that future research study teachers during problem-solving instruction, to determine whether some teachers are naturally more adept at such instruction than others. Unfortunately, in this project we did not have the time or resources to do this important exploratory work. Instead, we built a program based upon recommendations that were available in the literature and we integrated this advice with our own thinking.

We tested this new training manual in combination with a manual that had been developed in the previous experiment (in fourth-grade classrooms) in sixth-grade classrooms in the same school district. Pre- and posttests indicated that the program had a significant effect upon treatment students' verbal problem-solving skills. However, the program did not have a significant effect upon general mathematics achievement, probably because the general manual was commonly available in all schools (after the successful field experiment in grade four, all fourth-grade teachers in the district were given inservice training in the program). In addition to this possibility of general "contamination," comments made by some control teachers during debriefing procedures indicated that they were familiar with parts of the general program that are not routinely found in elementary school curricula (e.g., mental computations).

Finally, it should be noted that the success of the treatment in the second field experiment was moderated by the form of general administrative organization of mathematics teaching (math as special subject, semi-departmentalized, or open plans). The results of the program are therefore somewhat dependent upon teacher type, student type, and administrative organizations, as well as on the treatment per se.

The combined results of our work in elementary schools suggested that meaningful improvement in students' mathematics achievement was possible and that the programs which we had developed were reasonable intervention strategies, at least under certain conditions. With this experience, we moved into secondary schools.

The data collected in the junior high project indicate that change in teacher behavior and in student performance in secondary schools is possible. In particular, the results demonstrate that participation in the treatment program was associated with a significant positive effect upon students' problem-solving skills, as measured by the SRA test. Although the authors do not feel that the problems in this test are a completely adequate measure of problem solving, they do represent some skills that are important. It is therefore edifying to see that treatment teachers had a positive effect upon student performance on the problem-solving subtest.

Although we have raised some questions concerning the adequacy of the content criterion test and the level of teacher implementation, the overall evidence suggests that treatment teachers did implement more problem-solving strategies than did control teachers and that the test was a reasonable measure of content being presented in classrooms. We can thus confidently say that the program had a positive effect on treatment teachers' implementation and students' performance on the test.

We found that the opportunity to work with junior high teachers to modify our program was an interesting and valuable experience. In retrospect, we would have done some things differently (see Chapter 6). Still, we were reasonably pleased with the level of teacher involvement obtained and the ideas which were incorporated into the program. Several program modifications were made, and we think that these changes were appropriate and important for adapting the program to secondary schools. These ideas were essentially teacher initiated, and we are grateful to the participating teachers for their input and assistance. Teachers' brief involvement in training did appear to alter certain aspects of some teachers' behavior, and increased their involvement in the project. Furthermore, program implementation was found to have a significant and positive impact upon the problem-solving performance of junior high students.

However, project involvement did not have a positive effect on all teachers. These individual variations among teachers are similar to results reported by Ebmeier (1978) in the elementary school project. Some teachers in that study implemented the program more fully than others. In particular, Ebmeier found that program implementation was higher among teachers who felt that they were already teaching in ways recommended by the program, and by teachers who were searching for new alternatives. The brief partnership and the general training procedures which we utilized in this study would appear satisfactory for obtaining program implementation from such teachers. However, for teachers who are not interested in seeking alternative solutions and who feel that the program is contrary to their teaching styles, more elaborate procedures and more time will probably be necessary.

FUTURE RESEARCH

Working with Teachers

As noted in Chapter 6, there are a number of issues associated with working with teachers that merit future research consideration. Among the many issues that we can only speculate about at present are: How shall partnership teachers be selected? When and how shall "expert" external information be integrated with the "expert" local knowledge that teachers individually and collectively bring to a "partnership" with researchers? How many meetings are necessary and what intervals are useful for building rapport sufficient for direct and collaborative decision making? Under what conditions do teachers need the chance to observe and be observed by fellow teachers? We suspect that the answers to these (and other questions raised in Chapter 6) will vary with the context (the particular school district, the particular problem being investigated, and the types of generalizations which teachers and researchers are trying to make). More research is needed if we are to gain more skill in dealing with these issues.

STAFF DEVELOPMENT

Whether teachers are used by researchers as collaborators as active decision makers or not (i.e., in designing classroom interventions), there are countless issues surrounding implementation that merit research attention. As a case in point, the elementary school program we tested (see Chapter 3 and Appendix B) had tight time line specifications. Teachers, for example, were asked to try to teach development about 20 minutes each day. Thus, the positive results we obtained in that study were associated with a program that asked teachers to behave in reasonably standard ways across different curriculum topics.

We believed at the time we wrote the program that generally a balance between development and independent seatwork should be present in most lessons. However, we also knew that some content and some points in a lesson (introduction of concept vs. lessons in which concepts is being applied or reviewed) demand different ratio of development/seatwork. However, it is impossible to spell out all of these beliefs (let alone conduct research on each possible combination) about how time for development needs to be adjusted with content and the instructional cycle. Hence, we offered a generalized solution—a plan that would work much of the time.

A different approach to implementation would be to allow teachers more latitude in how they spend time on a particular day. We suspect that the results of such a treatment would be to produce a more variable set of findings. That is, some teachers would be excellent at applying general principles associated with active teaching to their particular students and goals. We suspect that other teachers would be less successful decision makers and that these variations would be less adequate than if they had implemented the average model presented in Chapter 3.

Teachers who are less successful than the average model may be poor planners or may tend to make decisions on the basis of convenience (today I will teach, tomorrow we will use the whole period for seatwork). Yet another reason we predict that some teachers will get poor results is because the adaptions for individual students involve very complex decisions. As Cronbach (1967) has written, "the poorer the differential information, the less the teacher should depart from the treatment that works best on the average . . . modifying treatment too much produces a worse result than teaching everyone alike" (p. 30). Information about how teachers modify the program in their teaching would be important, as would be more knowledge of criteria teachers utilize for making such decisions.

Among the many issues that can be examined profitably in the staff development area are (1) which teachers to include in a general instructional program—is it better to start only with volunteer teachers, and (2) how can teachers be encouraged to adopt general ideas from a program while maintaining their own stylistic preferences as teachers?

We have found in our own work that teachers' affective reactions to the program vary and although they are basically positive, there is sufficient justification to warrant serious thinking about how inservice training can be adapted to fit the needs of certain types of teachers. It would appear that certain teachers need fewer initial requests for changes in instructional behavior; other teachers appear able to handle many changes simultaneously. Much more research is needed on staff development/implementation variables.

We are still searching to find better ways to communicate our findings (e.g., we are developing films and new training procedures) and in particular we are trying to find ways to illustrate how program parts (e.g., controlled practice) can be adjusted to fit more appropriately with particular settings.

Development Lesson

In our work in elementary schools we found that many teachers did not regularly use an extended development component in their mathematics lessons. The Missouri Mathematics Program appeared to be helpful in elementary schools because it increased the amount of time elementary school teachers were utilizing for development, and it thus helped them to become more active in their teaching of mathematics. However, we found that most secondary teachers regularly included a development portion in their lessons and that time, per se, is not as important as is the quality of development. If improvements are to be made in teachers' instruction during development, it seems important to generate more adequate procedures for conveying to teachers criteria which can be used to estimate the quality of the development phases of their lessons. As a beginning step, we have tried to articulate our beliefs about development more clearly than we have in the past (see Chapter 9). Needed now is research to support and/or modify the present conceptualization.

More *content-focused development* needs to be emphasized in future research efforts. Although the program provides general strategies for teaching mathematics, particular content needs to be studied more thoroughly. Better conceptualization of the instructional demands of different types of mathematical content is needed and information about how the development portion of the lesson can be adjusted in ways that are consistent with changes in content. Although the purpose of the program is not to develop generic lesson plans, it is designed to encourage careful thinking and analysis by individual teachers. In its present form, the program does not do enough to promote critical thought about different types of mathematics content or about which strategies are more or less appropriate for teaching different types of content. Time allocations suggested in the program should probably vary with different types of mathematics content, as well as according to the lesson stage. The same sensitivity to variation should be built into the observational coding system and checklists which are used for classroom observation. Both the program's instructional strategies and observational procedures need to be more closely related to content issues in the teaching of mathematics.

Need to Go beyond a General Model

The model we have presented is an average or generalized approach to teaching mathematics. It is now time to explore instructional variations more fully, particularly in terms of the reciprocal effects of specific mathematics content being presented. Specific parts of the Missouri model and their effects upon particular types of teachers and students in the classroom need to be explained. Previously (in Chapter 8) we called for more component testing (Slavin, 1982) to determine if all parts of the model are necessary. Here we suggest that it will be important not to do component testing (e.g., is homework necessary) in the absence of specific types of content and learners.

The collective results of studies in our research program as well as those obtained elsewhere (e.g., Stallings, 1980; Anderson, Evertson and Brophy, 1979; Evertson, Anderson, Anderson, and Brophy, 1980) provide evidence that general training programs can have impact upon mean classroom performance. However, there are sufficient data to suggest that general treatment programs are apt to have different levels of impact on different combinations of teacher types and student types (Janicki and Peterson, 1981; Ebmeier and Good, 1979). Such results call for a need both to understand why programs affect different combinations of teachers and students in different ways and to develop procedures for developing more differentiated instructional programs.

We are pleased that our training program has had some success. However, it is important to reiterate that different teachers, various organizational structures, and diverse types of students have interacted in various ways with the program to affect the pattern of results. Much more information is needed about how these context conditions influence the program and the ways in which the general strategies and structures can be calibrated to fit into particular contexts.

Theory

Past research has shown that teachers vary in their behavior and in their effects on students. It is now time to synthesize the findings from our research program and research elsewhere, and to identify models for studying particular contexts. We must also learn how to adapt mathematics lessons to individual students and to particular types of content. Such a synthesis of empirical studies will be complex. To guide the synthesis of present results, to direct future research, and to gain insights into mathematics teaching, we will need to develop new theoretical constructs.

Since the Missouri Mathematics Program focused on whole-class instruction, it is difficult to speculate about its effects on particular learners or for particular content. Nevertheless, it might be instructive to present some hypothetical comments about why the Missouri Mathematics Program has appeared to work at a general level. These ideas have not yet been tested, but hopefully will be topics of future research.[1] The following comments are taken from Good (in press).

> We have evidence that the Missouri Mathematics Program in general had positive impact upon the *mean* performance of students in experimental classrooms, but we have no data to explain why the program worked. I suspect that the program had an impact because many elementary school teachers simply do not emphasize the meaning of the mathematical concepts they present to students, and they do not actively teach these concepts. Too much mathematics work in elementary schools involves some brief teacher presentation and a long period of seatwork. Such brief explanations for seatwork do not allow for meaningful and successful practice of concepts that have been taught, and the conditions necessary for students to discover or use principles on their own are also lacking.
>
> It seems plausible that the emphasis in our program upon the development stage of the lesson leads teachers to think more deeply about the concepts that they are presenting and to search more actively for better ways of presenting those concepts to students. Furthermore, given the way in which the development stage of the lesson is conducted, the program of instruction should allow teachers to see students' errors before they have a chance to practice those mistakes for a long period of time. This feature of the program seems to be especially desirable because some research has suggested that it is very difficult for students to tell teachers that they do not understand instruction. The clear development lesson would help students to understand more fully the concepts that they must master and how those concepts are related to other concepts that they have learned. The development phase of the lesson thus helps both teachers and students to develop a better rationale for learning activities and to develop a sense of continuity.
>
> The controlled practice portion of the lesson aids both teachers and students in understanding whether the basic concepts and mechanics are being understood. This is especially the case if teachers have developed the expectation that initial teaching often is associated with less than adequate student comprehension and that student mistakes call for reteaching, not rationalization. Such information especially allows teachers to correct and to reteach aspects of the lesson so that students develop appropriate conceptual understandings that students would be much more active thinkers during the development and controlled practice portions of the lesson. This is because students know that seatwork and their homework are intimately related to these activities. Hence, successful understanding during controlled practice leads to successful seatwork and successful homework. Checking of seatwork allows teachers one final opportunity to correct misunderstandings prior to the assignment of homework. Following successful practice, brief homework assignments should offer

---

1. There is some recent research evidence to illustrate that students are more attentive in treatment than control classes (see Harre, 1980).

students positive learning experiences that both provide for better integration of material and also the development of more appropriate student attitudes about mathematics and their ability to learn it. In particular, students will probably conclude that increased personal effort during mathematics instruction leads to positive learning experiences. Students would thus be presenting more positive feedback to teachers about mathematics instruction (e.g., handing in completed homework and exhibiting positive verbal and non-verbal behaviors during mathematics instruction, which in turn increases teachers' expectations that they can present mathematics effectively, leading to renewed efforts on their part to carefully structure the mathematics lesson).

The preceding statements are only a few of the beliefs and hypotheses that we hold about *why* the mathematics program was working. It is important to note that these hypotheses need to be tested if we are to develop more adequate understanding of the antecedent conditions necessary for successful mathematics learning. For example, research is needed to determine if in fact experimental teachers identify more student errors and can more readily understand those mistakes during the development stage than do control teachers who use different teaching techniques. It would be equally important to determine whether students in experimental classrooms are more active thinkers during the development portion of the lesson than are students in control classrooms (perhaps by asking students to do problems immediately after the development portion of the lesson). Similarly, more research is needed concerning the conditions under which student errors are developmentally helpful and lead to increased student effort to integrate material, rather than debilitating and convincing students that they do not understand mathematics. When teaching effectiveness studies begin to examine their embedded assumptions by stating and testing the specific ways in which student learning is influenced, the conditions under which teaching and learning strategies are useful will become clearer than they are at present.

Clearly, these comments are meant to explain why the mathematics program was working in elementary schools. We need to consider the contextual differences between teaching in elementary schools and teaching in secondary schools, to include such contextual differentiations in our theoretical thinking, and to begin to test these theoretical notions. Investigators must consider the perspectives of teachers, students, and researchers as they formulate theories and design studies. Historically, research on teaching has tended to emphasize one set of variables at the expense of others. For example, sometimes detailed clinical interviews are conducted with students, but classroom observations of those students are not made, and teacher opinions about what is taking place during instruction are not measured. If we are to increase understandings of classroom learning, it will be necessary to incorporate the more immediate responses of teachers and students into the design of classroom research (see, for example, Anderson, 1981). Researchers need to integrate subject matter variables (what is good "development" for a lesson on fractions vs. a multiplication concept). Some attempts to integrate the study of subject matter variables with more immediate classroom process are underway (e.g., Confrey and Lanier, 1980; Confrey and Good, in progress). Also, if theory is to be developed we must consider more fully the role that students can play as active teachers (see for example Webb, 1982; Peterson and Wilkinson, 1982) as well as the structural constraints that limit the effectiveness of teachers and students alike (e.g., Doyle, 1979; Bossert, 1979; Cohen, 1982).

Integrating Research Paradigms

Although our research began with an attempt to identify what effective teachers did in the classroom, we do not believe that this is necessarily the best way to understand teaching. We felt at the time that little was known about what takes place during instruction in elementary school mathematics and we wanted to understand the phenomenon more fully.

We used a quantitative procedure for this purpose, and we think that such large-scale research may be useful for identifying interesting sources of naturalistic variance (e.g., teacher behavior, student behavior, student outcomes, etc.) which may be worth investigation. For example, at present mathematics educators are very interested in problem-solving behavior and "solutions" for improving instruction are frequently offered, yet we have no information about what teachers do when they are teaching problem solving. The current focus on the learner in problem solving research must be supplemented by research on the teacher's role in developing problem solving ability (Grouws, 1981).

We know from our own data that some teachers who group students for instruction in mathematics have very positive effects on students, as measured by standardized achievement tests. Because our research design focused upon the extreme performers in our sample (most of whom happened to be whole-class or large-group teachers), we cannot describe how more and less effective small-group teachers varied in their behavior. There are probably many other important questions that can be addressed effectively by quantitative/naturalistic procedures. However, experiences and outcomes that many would like to see occurring in schools, but which do not presently exist in natural practice, would require changing behavior, not merely studying it naturalistically (Good, 1981).

Quantitative strategies are useful for identifying potentially interesting classroom practices (although in some cases the findings obtained by such strategies may be misleading), and for describing such practices in rough outlines. However, quantitative strategies seem to be a poor methodology for explaining classroom patterns. Furthermore, the method by which quantitative data are collected may encourage researchers to ask the wrong types of questions. For example, for a long time researchers have studied (without much success) the relationship between cognitive levels of questions that teachers ask student and students' subsequent achievement. However, a much more important factor than cognitive levels of questions per se may be the pattern of meaning that a set of questions creates. Researchers might also try to determine whether a set of questions (e.g., across a mathematics lesson) is a deliberate teacher attempt to present a concept and to measure students' comprehension, or if the questions appear to be unrelated.

This topic of teachers' questions is only one of many which could be used to illustrate the need for more rigorous studies of class behavior, including information about participants' perceptions of those behaviors (why do teachers believe they ask questions and what do they feel such behaviors communicate to students, and reciprocally how do students interpret teachers' questions).

In our initial naturalistic study we believed that teachers who were more effective in obtaining student achievement were much more likely to question students in a diagnostic cycle. That is, effective teachers present concepts and information as clearly as possible and then ask students relatively simple questions in order to measure their basic understanding. Teachers can thus assure themselves that students have comprehended the material and then teachers can deal with process and more complex issues. Researchers interested in theory and in the design of instruction to fit a particular context will have to explain dynamic patterns of pupil and teacher responses and their mutual adaptation. Qualitative strategies are better for addressing such issues as social language in learning, the subjective interpretations of teachers about students, negotiated meaning, and reciprocal effects of classroom participants (see for example Anderson-Levitt, 1981; Levine and Mann, 1981; Florio, 1979; Erickson, 1973).

Qualitative strategies offer no guarantee of increased understanding of classrooms. All too often in qualitative research investigators study only a few aspects of the environment in great detail while other variables potentially affecting classrooms are ignored. If new theories and more differentiated instructional programs are to emerge, more comprehensive and integrative strategies will have to be utilized. In new research efforts, quantitative strategies may be used to associate problems with an appropriate sample. For example, an investigator interested in the polarization of high- and low-achieving students in American classrooms might want to use quantitative strategies to determine whether there are classrooms where low-achieving students do not fall further behind high-achieving students as the year progresses. If such classrooms exist, they would provide an interesting contrast to classrooms in which increased polarization does occur. After potential samples are identified (and quantitative strategies would often be useful for this purpose), then quantitative and qualitative methodologies could be used in combination to explore the problem. We are clearly advocating that researchers match their research strategy to the problem being studied. We are also suggesting that research become more integrative. It seems unfortunate that research focuses upon social learning *or* academic learning; upon students *or* teachers; or upon management *or* instruction.

A potentially useful approach at this time would be a more comprehensive study of the major aspects of classrooms simultaneously (teacher perception and behavior; student perception and behavior; curriculum content; social learning). As the scope and breadth of studies increase, it may also be advantageous to increase the range of competencies that individuals bring to the research task. Individual experts can work on various aspects of a large study. Obviously, there are limits to the number of variables which can be included in a single study and only certain aspects of classrooms can be examined, even with large amounts of time and money. There are also problems in securing and maintaining working relationships in cross-discipline research teams. However, despite these problems, it seems important for researchers to study more classroom variables than they have in the past. One means by which such expansion of research might take place is a cross-discipline research team whose individual members share a commitment to a common research problem, even though they have different methodological and substantive skills and insights.

SUMMARY

We believe that the research discussed in this book provides strong evidence that elementary and secondary teachers can and do make a difference in student learning in mathematics. The data illustrate that teacher effects are sufficiently large as to have consequences for educational and social policy. Specifically, we believe that teachers represent a distinct educational resource and that money invested in attracting, educating, and retaining capable teachers can result in increased student achievement.

Data presented here also suggest that teachers can be trained (in ways that are relatively inexpensive) to teach in new and improved ways that positively influence students' achievement. Furthermore, it is our belief that teachers can become more organized and place increased attention upon the meaning of mathematical concepts and at the same time maintain a relaxed but task-oriented classroom environment for their students.

We stress the quality of teaching and the meaningfulness of the mathematics content presented because of the present focus in the literature on time utilization. We agree that students do learn more when they are on task longer. However, the important factor is the appropriateness of the task and whether students comprehend and integrate information and concepts in ways that will transfer to actual settings. We feel that some teachers are more effective than others in presenting mathematics in ways that increase students' comprehension and functional use of mathematical ideas, and that continued research efforts to learn from successful practices are warranted.

In this book we have discussed the Missouri Mathematics Program we have developed and tested in several studies. Our work, as well as that conducted with the program by others, suggests that the materials are useful for inservice training and can lead to improved teacher and student behavior. Although we are pleased that the program has had some positive influence, we do not intend to argue that teachers should necessarily use this program. However, we believe that teachers who present a careful development lesson, who look for signs of student comprehension (and who reteach lessons when students have difficulty), and who provide students a chance to apply mathematical ideas by building into their teaching systematic review, practice, and process feedback will obtain better achievement than teachers who do not. There are many ways in which teachers can incorporate such principles into their teaching and our plan is only one way of doing so.

To reiterate, we believe our major contribution is the documentation of a clear teacher effect on student learning. Also, we have developed a few beliefs about why and how teachers affect student achievement and have presented this perspective here. Perhaps more importantly, we have discussed some of the weaknesses of our research and have indicated the major directions that such research should take in the future. We believe that teaching is partly art and partly science, and always will be. However, the systemic study of classroom learning and teaching does yield concepts and ways of viewing the practice of teaching that can assist in the development of teaching strategies which are better suited to the learning needs of students. In this regard, we hope that this book will encourage readers to think more systematically about instructional issues.

Appendix A

Measurement Instruments for Assessing Students'

(Aptitude/Attitude Inventory) and Teachers'

(Teaching Style Inventory) Beliefs About Mathematics

A. Aptitude/Attitude Inventory

ATTITUDE INVENTORY

Name _____

Boy _____Girl _____

Teacher's Name _____

School's Name _____

Directions:

    Read each statement and decide if you usually agree or disagree with that statement. If you agree, circle the letter T for True next to the question. If you disagree, circle the letter F for False next to the question.

    Please answer every question. Be sure you write your name, your sex, your teacher's name, and your school's name on this sheet. If you have a question, ask your teacher for help.

T F 1. I like to work my math problems with several other students.

T F 2. I always like to choose what math problems to do.

T F 3. I get into trouble in school about once every week.

T F 4. I do not like to work alone.

T F 5. I work harder on math problems that I know will be checked.

T F 6. I need to learn math.

T F 7. I need to be reminded often to get my math assignment done.

T F 8. I want to get good math grades just to show my friends.

T F 9. I sometimes forget to do my assignments.

T F 10. Practicing new math problems with my teacher is a waste of time.

T F 11. I do not need any practice work before I start work on new math problems.

T F 12. I can always remember what I am told to do.

T F 13. I usually finish the easy math problems but not the hard ones.

T F 14. I like my teacher to work a few example problems before I have to do a new problem by myself.

T F 15. I like to learn about math best by listening to my teacher.

T F 16. I will get good math grades this year.

T F 17. I am not good at math games.

T F 18. I usually finish my math assignments.

T F 19. I am good at working math problems in my head.

T F 20. I get into trouble in school about once every week.

T F 21. I like to do math problems in my own way.

T F 22. My teacher really wants me to get good grades in math.

T F 23. I usually do not finish my math assignment.

T F 24. Getting good grades in math is really important to me.

T F 25. I am good at working math problems in my head.

T F 26. I sometimes lose my books and papers.

T F 27. I like to have my parents help me with my math problems.

T F 28. I like to work math problems by myself.

T F 29. I like to learn about math best by reading my book.

T F 30. I always like to choose what math problems to do.

TURN THE PAGE OVER

T F 31. I like to figure out how to work a new math problem without my teacher's help.

T F 32. I will need math next year.

T F 33. Before I start working new math problems, I like to make sure I can do them.

T F 34. I like to learn about math best by listening to my teacher.

T F 35. I do not like to check my math problems.

T F 36. I like to know if a math assignment will be checked.

T F 37. It is not that important to know math.

T F 38. If I have a question in my math class, I ask the teacher right away.

T F 39. Other subjects are more important than math.

T F 40. My math teacher last year yelled at me a lot.

T F 41. I want to get good grades just for myself.

T F 42. If I find out why I made a mistake on a math problem, I usually do not miss that kind of problem again.

T F 43. I like to be able to choose what our class does in math.

T F 44. I like to have my teacher explain how to work a new math problem.

T F 45. I will get good math grades this year.

T F 46. I do not like to check my math problems.

T F 47. Getting good grades in math is really important to me.

T F 48. If I know my math problems will not be checked, I do not work on them very much.

T F 49. I like to check my math problems to see which problems I missed.

T F 50. I work harder if I know my math problems will be checked.

T F 51. I like to work math problems in my head.

Answer the following questions by circling . . .

1 if you want to answer <u>always</u>
2 if you want to answer <u>most of the time</u>
3 if you want to answer <u>sometimes</u>
4 if you want to answer <u>never</u>

Always
Most of the time
Sometimes
Never

1 2 3 4 52. Do you like to be in this class?

1 2 3 4 53. Do you have much fun in this class?

1 2 3 4 54. Do most of your close friends like the teacher?

1 2 3 4 55. Does the teacher help you enough?

1 2 3 4 56. Do you learn a lot in this class?

1 2 3 4 57. Do you ever feel like staying away from this class?

1 2 3 4 58. Are you proud to be in this class?

1 2 3 4 59. Do you always do your best in this class?

1 2 3 4 60. Do you talk in class discussions in this class?

1 2.3 4 61. Are most of the students in this class friendly to you?

B. Teaching Style Inventory

Name _____

School _____

Part I CLASSROOM PROCEDURES

Please check the point within each of the following scales which most accurately describes your math class. (If you are teaching math for the first time or your present situation is very different from previous years, please respond as you antici-pate your class will be like this year.) Please respond accord-ing to what actually happens, not what you think should happen, or what you would like to have happen. There are no right or wrong answers. Please answer all the questions.

1. Amount of testing

   I give a math test about once every three weeks. [1]
   [2]
   [3]
   [4]
   I give a test at least once every week. [5]

2. Emphasis on enjoyment

   Very strong explicit emphasis is put on having a pleasant, happy and friendly time in my math class. [1]
   [2]
   [3]
   Although having an enjoyable time in math is important there is little explicit emphasis on having a pleasant, happy and friendly time in my math class. [4]
   [5]

3. Task emphasis

   The importance of getting work done on time and done well is frequently stressed in my class. [1]
   [2]
   [3]
   [4]
   Students can turn in their work when they are finished. There are no strict deadlines. [5]

4. Organization of tasks

   Most learning tasks in this class have a step-by-step organization and sequence. [1]
   [2]
   [3]
   [4]
   Most of the learning tasks in this class are "open-ended" or discovery oriented. [5]

5. Commonality

   Math learning objectives are the same for all students in the class. [1]
   [2]
   [3]
   [4]
   Math learning objectives are set for each student separately. [5]

6. Problems

   Students are encouraged to get a lot of help with their math problems. [1]
   [2]
   [3]
   [4]
   Students are encouraged to solve their math problems without a lot of teacher help. [5]

7. Help with work

   Almost all help is initiated by students asking for it. [1]
   [2]
   [3]
   [4]
   Almost all help is initiated by my seeing the need for it. [5]

8. Plan changing

Daily lesson plans are stable, not usually subject to change. |__| 1

|__| 2

|__| 3

|__| 4

Daily lesson plans are changed very frequently. |__| 5

9. Different activities

Many different activities are almost always going on simultaneously during math class. |__| 1

|__| 2

|__| 3

|__| 4

Almost all the time the students are all engaged in the same activity during math class. |__| 5

10. Evaluation standards

The same standards are used for all students. |__| 1

|__| 2

|__| 3

|__| 4

Different standards are used for each individual. |__| 5

11. Evaluation procedures

Evaluation procedures are the same for all students in the class. |__| 1

|__| 2

|__| 3

|__| 4

Evaluation procedures are different for each student. |__| 5

12. Oral presentation

On a typical day, I give an oral presentation for three-fourths of the math time. |__| 1

|__| 2

|__| 3

|__| 4

I almost never give an oral math presentation. |__| 5

13. Peer help

Students frequently help one another during math class. |__| 1

|__| 2

|__| 3

|__| 4

Students seldom help one another during math class. |__| 5

14. Instructional direction

On a typical day, I direct my attention to the math class as a group three-fourths of the time or more. |__| 1

|__| 2

|__| 3

|__| 4

On a typical day, I teach or direct my attention to individual students (or small groups) three-fourths of the time or more. |__| 5

15. Approaches to learning

I encourage students to solve a given math problem the way I have demonstrated. |__| 1

|__| 2

|__| 3

|__| 4

I encourage students to solve math problems any way that they desire. |__| 5

16. Conceptualization

I use conceptual ideas, such as the commutative and associative properties of addition and multiplication to teach math. |__| 1

|__| 2

|__| 3

|__| 4

I teach math from a more practical, less theoretical point of view. |__| 5

17. Inductive-deductive approach

    I present a math concept first then illustrate that concept by working several problems (deductive).

    ___1 ___2 ___3 ___4 ___5

    I present the class with a series of similar problems, then together we develop concepts and methods of solving the problems (inductive).

18. Curriculum organization

    The curriculum is organized such that certain topics are repeated (but in more depth) on a regular basis throughout the year.

    ___1 ___2 ___3 ___4 ___5

    Once a certain topic is covered, that same topic is not covered again except during reviews.

19. Transfer

    A good deal of the time (1/3) is spent trying to teach students to see similarities and differences between new and previously learned math ideas.

    ___1 ___2 ___3 ___4 ___5

    New topics are generally introduced with limited reference to previously learned math ideas.

20. Practicality

    Math is taught strictly as a practical subject.

    ___1 ___2 ___3 ___4 ___5

    Math is taught with emphasis on theory.

21. Predictability of student pace

    I can usually predict where my students will be in the math textbook in January.

    ___1 ___2 ___3 ___4 ___5

    I can't usually predict where my students will be in the math textbook in January.

22. Student choice

    Students have a choice as to what problems or exercises they can do for math practice.

    ___1 ___2 ___3 ___4 ___5

    I decide what problems the students will do for math practice.

23. Pre-assessment

    I know a good deal about my students' math abilities before or shortly after the school year starts.

    ___1 ___2 ___3 ___4 ___5

    It usually takes about 9 weeks before I know about my students' math abilities.

24. Motivation

    All students are rewarded in the same manner for good work.

    ___1 ___2 ___3 ___4 ___5

    Students are rewarded in different ways for good work.

25. Mobility

    Students seldom stay in their seats for the major part of the math lesson.

    ___1 ___2 ___3 ___4 ___5

    Students are generally in the same seat for the math period.

26. Math emphasis

    In my math class I emphasize the basic computational skills.

    ___1 ___2 ___3 ___4 ___5

    In my math class I emphasize understanding the concepts underlying mathematics.

27. Study places

    Each child works mostly at his own desk during math lesson.

    ___1 ___2 ___3 ___4 ___5

    All math work is divided among a variety of places (centers) in and out of the classroom, with no "home base" seat.

28. Instructional changes

    I seldom change my approach throughout the semester (such as lecture-discussion, discovery, etc.).

    ___1 ___2 ___3 ___4 ___5

    I change my approach frequently (from discovery to direct telling or from another method to something different) throughout the semester.

29. Changes

    The arrangement of furniture and equipment has changed every week or so, this year.

    ___1 ___2 ___3 ___4 ___5

    The arrangement has changed once or not at all.

30. Rule enforcement

    I enforce the classroom rules.

    ___1 ___2 ___3 ___4 ___5

    Students enforce classroom rules.

31. Rule making

    I make the classroom rules.

    ___1 ___2 ___3 ___4 ___5

    Students make the classroom rules.

32. Reinforcement

    I generally use concrete reinforcers such as stars.

    ___1 ___2 ___3 ___4 ___5

    I generally use verbal praise as reinforcement.

33. Affective objectives

    Appreciation of math is of high importance.

    ___1 ___2 ___3 ___4 ___5

    Appreciation of math is not vital.

34. Emphasis on consumer math

    Heavy emphasis is placed on consumer math.

    ___1 ___2 ___3 ___4 ___5

    Little emphasis is placed on consumer math.

Part II TEACHER OPINION

Select the appropriate choice for each statement.

A = Agree
B = Somewhat agree
C = Undecided
D = Somewhat disagree
E = Disagree

40. ___ Teaching math makes me feel secure and at the same time it is stimulating.

41. ___ Teaching multiplication and division is more enjoyable than teaching geometry or fractions.

42. ___ In terms of teaching skill, math, in comparison to other subjects and activities I teach, is a personal strength.

43. ___ Math, in comparison to other subjects and activities I direct, is one of my lesser interests.

44. ___ Math is one of the few areas in which poor readers can do well.

45. ___ My basic function as a math teacher is to convey my knowledge of math to the students in a direct manner.

46. ___ Boys in my class have more interest in math than girls do.

47. ___ Without the assistance of a special teacher (i.e., a specialist in mathematics), the classroom teacher should not be regarded as responsible for the limited progress made by the slowest pupil.

48. ___ Individualization of math instruction seems impractical for actual classroom application.

49. ___ If resources were available, I would prefer total individualization of math instruction rather than group or whole class instruction.

50. ___ I feel I have a good sound background in mathematics.

35. Sex differences

Boys are better in math skills.

Girls are better in math skills.

36. Divergence from planned lesson

I try hard to stick to the lesson planned for that day during math period.

If a student raises an interesting question during the math lesson, I may change my whole lesson plan for that day and pursue the student's question.

37. Emphasis on comprehension

Understanding the methodology of why a given method gives the correct answer is important.

Understanding the methodology is not critical.

38. Exploration

Most of the time is spent drilling the student in math fundamentals.

Most of the time is spent exploring math-related topics.

39. Pacing

Most math class activities require students to work at about the same pace; topics are expected to be mastered by specific times during the year.

Each student works at his or her own pace, with no timing restrictions.

Part III COMPLETION

51. Most of my students complete ___ % or more of all the problems in their textbook associated with each lesson that is taught.

52. As of today, I have ___ students that are discipline problems.

53. When you use practice exercises to reinforce math skills, approximately what percentage are:

    ___ written work to be done in class

    ___ written work to be done at home

    ___ oral work or chalkboard work

    ___ games or puzzles that illustrate the concept

    ___ other

    100%

54. When some students do poorly on tests or otherwise indicate that they have not understood a unit in math, what are three (3) things you do to improve the situation.

    A. _____

    B. _____

    C. _____

55. On the average I spend about ___ minutes a day developing math concepts and skills and have the children practice these skills through homework and problems ___ minutes a day.

56. This year I teach math ___ days a week for an average of ___ minutes a day.

57. My students should have the opportunity to select and use math materials on a nonstructured basis at least ___ times a week.

58. I assign math work to be done at home about ___ times a week.

59. Sometimes students have difficulty solving story problems. Briefly describe how you help your students solve story problems. (Example: I have pupils make drawings or diagrams to help clarify the problem.)

60. When you correct students' papers, how would you describe the type of marks you most often put on the students' papers? (Example: I mark the problems that are incorrect and provide the correct answer.)

61. How often do you review material already covered? (Example: At the end of the chapter, before vacations, etc.)

62. When I assign students math story problems, I go over the vocabulary in the problem and point out what new words mean about ___ % of the time.

63. Before I start presenting the math lesson for the day, I spend about ___ minutes going over the previous lesson.

64. The students in my class make use of or manipulate concrete educational equipment (such as blocks, compasses, rulers, etc.) to aid in understanding math concepts about ___ times a week.

65. I move the students into new material when I feel that all but about ___ % of the students are ready.

66. During the year when you start a new math unit that is especially difficult, what do you do differently? (Example: I present the material more slowly than normal and I assure the students they can handle the new material.)

67. Given my present objective and methods of teaching, I feel the ideal class size in math would be ___ (number) students and that the maximum number I could teach and still do a good job would be ___ (number) students.

68. How many years (including this year) have you taught math to fourth grade students?

    ___ years

69. How many years (including this year) have you taught in an elementary school setting?

    ___ years

70. How many hours of college credit in math have you completed (including math methods courses)?

    ___ hours

71. How many hours of graduate college credit (including courses you may presently be enrolled in) have you completed beyond the B.A. or B.S. degree?

    ___ hours

72. When math assignments are checked, what percentage would fall into the following categories?

    _____ I check the students' papers.

    _____ An aid checks the students' papers.

    _____ Students check their own work.

    _____ Students check each other's work.

    _____
    100%

73. If you had your choice, what type of ability in math would you prefer to teach? (Check one.)

    _____ mostly high ability

    _____ mostly average ability

    _____ mostly low ability

    _____ a mixture of abilities

References

Anderson, L. Short term responses to classroom instruction. *Elementary School Journal*, 1981, *82*(2).

Anderson, L., Evertson, C., and Brophy, J. An experimental study of effective teaching in first-grade reading groups. *Elementary School Journal*, 1979, *19*, 193–223.

Anderson, L., and Liu, J. Concerns for stability of aptitudes within the aptitude treatment interaction model. Paper presented at the annual meeting of the American Educational Research Association, San Francisco, April 1979.

Anderson-Levitt, K. Memory and talk in teachers' interpretations of student behavior. Paper presented at the annual meeting of the American Educational Research Association, Los Angeles, 1981.

Andros, K., and Freeman, B. The effects of three kinds of feedback on math teaching performance. Paper presented at the annual meeting of the American Educational Research Association, Los Angeles, 1980.

Baird, L. Teaching styles: An exploratory study of dimensions and effects. *Journal of Educational Psychology*, 1973, *64*, 15–21.

Beckerman, T. A study of the main and interactive effects of student types, sex, and treatment program on the mathematics achievement and attitudes of fourth grade students. Unpublished doctoral dissertation, University of Missouri, Columbia, 1981.

Behnke, G., et al. Coping with classroom distractions. *Elementary School Journal*, 1981, *81*, 135–155.

Bennett, N. *Teaching styles and pupil progress*. London: Open Book, 1976.

Berliner, D., and Cahen, L. S. Trait-treatment interaction and learning. In F.N. Kerlinger (Ed.), *Review of research in education, 1*. Itasca, Illinois: Peacock, 1973.

Berliner, D., and Tikunoff, W. The California beginning teacher evaluation study: Overview of the ethnographic study. *Journal of Teacher Education*, 1976, *27*, 24–30.

Biber, B. Teacher education in mental health—from the point of view of the psychiatrist. In M. Krugman (Ed.), *Orthopsychiatry and the school*. New York: American Orthopsychiatric Association Inc., 1958.

Bloom, B. *Human characteristics and school learning*. New York: McGraw-Hill, 1976.

Bossert, S. Task and social relationships in classrooms: A study of classroom organization and its consequences. American Sociological Association, *Arnold and Caroline Monograph Series*. New York: Cambridge University Press, 1979.

Bracht, G. Experimental factors related to aptitude-treatment interactions. *Review of Educational Research*, 1970, *40*, 627–645.

Britt, D. An improved method for instructional development: Learner types. *Audiovisual Instruction*, 1971, *16*(4), 14–15.

Brophy, J. Teacher behavior and its effects. *Journal of Teacher Education*, 1971, *71*, 733–750.

Brophy, J. How teachers influence what is taught and learned in classrooms. *Elementary School Journal*, 1982 (in press).

Brophy, J., and Evertson, C. Process-product correlations in the Texas teacher effectiveness study: Final Report (RES. REP. 74–4). Austin, Texas: R & D Center for Teacher Education, 1974. (ERIC No. ED 091 394).

Brophy, J., and Evertson, C. *Learning from teaching: A developmental perspective*. Boston: Allyn and Bacon, 1976.

Brophy, J., and Good, T. Dyadic system. In A. Simon and E. Boyer (Eds.), *Mirrors for behavior: An analogy of observation instruments*. Philadelphia: Research for Better Schools, Inc., 1970.

Brophy, J., and Good, T. *Teacher-student relationships: Causes and consequences*. New York: Holt, Rinehart and Winston, 1974.

Broverman, D. Cognitive styles and intra-individual variation in abilities. *Journal of Personality*, 1960, *28*, 240–256.

Bush, R. *The teacher-pupil relationship*. Englewood Cliffs, New Jersey: Prentice-Hall, 1954.

Canfield, A., and Lafferty, J. *Learning styles inventory*. Detroit: Humanities Media (Liberty Drawer), 1970.

Carroll, J. A model of school learning. *Teachers' College Record*, 1963, *64*, 723–733.

Clement, J. Algebra word problem solutions: Thought processes underlying a common misconception. *Journal for Research in Mathematics Education*, 1982, *13*, 31–49.

Cohen, E., and Anthony, B. Expectation states vary in classroom learning. Paper presented at the annual meeting of the American Educational Research Association, New York City, 1982.

Coleman, J. Methods and results in the IEA studies of effects of school on learning. *Review of Educational Research*, 1975, *45*, 355–386.

Coleman, J., Campbell, E., Hobson, C., McPartland, J., Mood, A., Weinfield, F., and York, R. *Equality of educational opportunity*. Washington, D.C.: Superintendent of Documents, U.S. Government Printing Office, 1966.

Confrey, J., and Good, T. A view from the back of the classroom: Integrative student and teacher perspectives of content of observational and clinical interviews, in progress.

Confrey, J., and Lanier, P. Students' mathematics abilities: Improving the teaching of general mathematics. *School Science and Mathematics*, 1980, *80*, 549–556.

Coop, R., and Sigel, I. Cognitive style implications for learning and instruction. *Psychology in the Schools*, 1971, *8*, 152–161.

Cooper, H. Pygmalion grows up: A model for teacher expectation communication and performance influence. *Review of Educational Research*, 1979, *49*, 3, 389–410.

Crist, J., and Hawley, B. Individual differences and mathematics achievement: An investigation of aptitude-treatment-interaction in an evaluation of three instructional approaches. Washington, D.C. (National Institute of Education, OEC–0–71–4751), 1976. (ERIC No. ED 013 371).

Cronbach, L. How can instruction be adapted to individual differences? In R. Gage (Ed.), *Learning and individual differences*. Columbus, Ohio: C. E. Merrill, 1967.

Cronbach, L., and Snow, R. *Aptitudes and instructional methods*. New York: Irving/Naiburg, 1977.

Cunningham, W. The impact of student-teacher pairings and teacher effectiveness. *American Educational Research Journal*, 1975, *12*(2), 169–189.

Doyle, W. Classroom tasks and students' abilities. In P. Peterson and H. Walberg (Eds.), *Research on teaching: Concepts, findings, and implications*. Berkeley, California: McCutchan Publishing Corp., 1979.

Dubriel, J. A study of two plans for utilization of class time in first-year algebra. Unpublished doctoral dissertation, University of Missouri, Columbia, 1977.

Dunkin, M., and Biddle, B. *The study of teaching*. New York: Holt, Rinehart and Winston, 1974.

Dunn, R., Dunn, K., and Price, G. *Learning Style Inventory*. Lawrence, Kans.: Price Systems, 1975.

Dunn, R., and Dunn, K. *Teaching students through their individual learning styles*. Reston, Virginia: Reston Publishing Co. 1978.

Dunn, R., et al. Learning style researchers define differences differently. *Educational Leadership*, 1981, *38*(5), 372–376.

Durkin, D. What classroom observations reveal about reading comprehension instruction. *Reading Research Quarterly*, 1979, *14*, 481–533.

Ebmeier, H. An investigation of the interactive effects among student types, teacher types and treatment types on the mathematics achievement of fourth grade students. Unpublished doctoral dissertation, University of Missouri, Columbia, August 1978.

Ebmeier, H., and Good, T. The effects of instructing teachers about good teaching on the mathematics achievement of fourth-grade students. *American Educational Research Journal*, 1979, *16*, 1–16.

Emmer, E. *Classroom observation scales*. Austin, Texas: Research and Development Center for Teacher Education, 1973.

Engelhardt, J. The effects of systematic instruction in verbal problem solving on the achievement of sixth-grade students. Unpublished doctoral dissertation, University of Missouri, Columbia, 1980.

Erickson, F. What makes school ethnography "ethnographic?" *Council on Anthropology and Education Newsletter*, 1973 *4*(2), 10–19.

Evertson, C., Anderson, C., Anderson, L., and Brophy, J. Relationships between classroom behaviors and student outcomes in junior high mathematics and English classes. *American Educational Research Journal*, 1980, *17*, 43–60.

Fischer, B., and Fischer, L. Styles in teaching and learning. *Educational Lead-*

ership, 1979, *36*(4), 245–254.

Florio, S. The problem of dead letters: Social perspectives on the teaching of writing. *Elementary School Journal*, 1979, *80*(1), 1–7.

Freeman, D., Kuhs, T., Porter, A., Knappen, L., Floden, R., Schmidt, W., and Schwille, J. The fourth-grade mathematics curriculum as inferred from text-books and tests. Research Series N. 82, Institute for Research on Teaching, Michigan State University, 1980.

Gage, N. Address appearing in "Proceedings," *Research Resume, Vol. 16*. Bur-lingame, California: California Teachers Association, 1960.

Gardner, R., Holzman, P., Klein, G., Linton, H., and Sponce, D. S. Cogni-tive contol: A study of individual consistencies in cognitive behavior. *Psycho-logical Issues*, 1959, *1*, Monograph No. 4.

Good, T. Teacher effectiveness in the elementary school: What we know about it now. *Journal of Teacher Education*, 1979, *30*, 52–64.

Good, T. Classroom expectations: Teacher-pupil interactions, In J. McMillan (Ed.), *The social psychology of school learning*. New York: Academic Press, 1980.

Good, T. A decade of research on teacher expectations. *Journal of Educational Leadership*, 1981, *38*, 415–423.

Good, T. Classroom research. In L. Shulman (Ed.), *Teaching and educational policy*. New York: Longman Inc., in press.

Good, T., and Beckerman, T. An examination of teachers' effects on high, mid-dle, and low aptitude students' performance on a standardized achievement test. *American Educational Research Journal*, 1978, *15*, 477–482.

Good, T., Biddle, B., and Brophy, J. *Teachers make a difference*. New York: Holt, Rinehart and Winston, 1975.

Good, T. and Brophy, J. Behavioral expression of teacher attitudes. *Journal of Educational Psychology*, 1972, *63*, 617–624.

Good, T., Ebmeier, H., and Beckerman, T. Teaching mathematics in high and low SES classrooms: An empirical comparison. *Journal of Teacher Edu-cation*, 1978, *29*, 85–90.

Good, T., and Grouws, D. Process-product relationships in fourth-grade mathematics classrooms. Final Report of National Institute of Education

Good, T., and Grouws, D. Teacher rapport: Some stability data. *Journal of Edu-cational Psychology*, 1975(a), *67*, 179–182.

Grant NIE–G–00–3–0123, University of Missouri, Columbia 1975(b).

Good, T., and Grouws, D. Teaching effects: Process-product study in fourth-grade mathematics classrooms. *Journal of Teacher Education*, 1977, *28*, 49–54.

Good, T., and Grouws, D. The Missouri mathematics project: An experimental study of fourth-grade classrooms. *Journal of Educational Psychology*, 1979, *71*, 355–362.

Good, T., and Grouws, D. Experimental research in secondary mathematics class-rooms: Working with teachers (NIE–G–79–0103 Final Report), May 1981.

Good, T. Grouws, D., Beckerman, T., Ebmeier, H., Flatt, L., and Schneeberger, S. *Teaching manual: Missouri mathematics effectiveness project*. Technical Report No. 99. Center for Research in Social Behavior, University of Mis-souri, Columbia, 1977.

Good, T., and Power, C. Designing successful classroom environments for different types of students. *Journal of Curriculum Studies*, 1976, *8*, 1–16.

Good, T., and Stipek, D. Individual differences in the classroom: Psychological perspective. In M. Fenstermacher and J. Goodlad, *1983 NSSE Yearbook*, in press.

Gregorie, A. Learning/teaching styles: Potent forces behind them. *Educational Leadership*, 1979, *36*(4), 234–236.

Grouws, D. "The Teacher Variable in Mathematics Instruction." In *Selected Issues in Mathematics Education*, M. Lindquist (Ed.), National Society for the Study of Education, Berkeley, CA: McCutchan, 1980.

Grouws, D., and Thomas, W. Problem solving: a panoramic approach. *School Science and Mathematics*, 1981, *81*, 307–314.

Grouws, D. An approach to improving teacher effectiveness. *Cambridge Journal of Education*, 1981, *11*, 2–14.

Guilford, J. Three faces of intellect. *American Psychologist*, 1959, *14*, 469–479.

Harre, R. An investigation of the interactive effects among student types and treatment types on the time-on-task behavior in eighth-grade mathematics classes. Unpublished doctoral dissertation, University of Missouri, Columbia, 1980.

Harvey, O. Conceptual systems and attitude change. In C. W. Sherif and M. Sherif (Eds.), *Attitude: Ego-involvement and change*. New York: 1967.

Harvey, O., Hunt, D., and Schroder, H. *Conceptual systems and personality organization*. New York: Wiley, 1961.

Heath, R., and Nielson, M. The research basis for performance-based teacher education. *Review of Educational Research*, 1974, *44*, 463–484.

Heil, L., Powel, M., and Feifer, I. Characteristics of teacher behavior and competency related to the achievement of different kinds of children in several elementary grades. (U.S. Department of Health, Education and Welfare, Office of Education, Cooperative Research Project No. 7285). Brooklyn: Brooklyn College, 1960.

Hill, J. *Personalizing educational programs utilizing cognitive style mapping*. Bloomfield Hills, Michigan: Oakland Community College, 1971.

Holzman, P., and Klein, G. Cognitive system-principles of leveling and sharpening: Individual differences in assimilation effects in visual time error. *Journal of Psychology*, 1954, *37*, 105–122.

Hulleman, H. Personal communication based upon a research report entitled "Improving math skills . . . An evaluation of the Missouri mathematics effectiveness project" by Jim Gard, Principal, Indian Creek Elementary School, July 1981.

Hunt, D. *Matching models in education: The coordination of teaching methods with student characteristics* (Monograph Series No. 10). Toronto, Canada: Ontario Institute for Studies in Education, 1971.

Jackson, P., Silberman, M., and Wolfson, B. Signs of personal involvement in teachers' descriptions of their students. *Journal of Educational Psychology*, 1969, *60*, 22–27.

Janicki, T., and Peterson, P. Aptitude-treatment interaction effects of variations in direct instruction. *American Educational Research Journal*, 1981, *18*(1), 63–82.

Kagan, J., Moss, H., and Sigel, I. The psychological significance of styles of conceptualization. In J. F. Wright and J. Kagan (Eds.), *Basic cognitive processes in children*. Monograph of the Society for Research in Child Development, 1963, *28*, 93–112.

Kagan, N. Educational implications of cognitive styles. In G. Lesser (Ed.), *Psychology and educational practice*. Glenview, Illinois: Scott, Foresman and Co., 1972.

Keefe, J. (Ed.). *Student learning styles: Diagnosis and prescribing programs*. Reston, Virginia: National Association of Secondary School Principals, 1979.

Keziah, R. Implementing instructional behaviors that make a difference. *Centroid* (North Carolinal Council of Teachers of Mathematics), 1980, *6*, 2–4.

Kolb, D. Disciplinary inquiry norms and student learning styles: Diverse pathways for growth. In A. Chickering (Ed.), *In the modern American college*. San Francisco: Jossey-Bass, 1981.

Kounin, J. *Discipline and group management in classrooms*. New York: Holt, Rinehart and Winston, 1970.

Kozol, J. *Death at an early age*. Boston: Houghton Mifflin, 1967.

Kulm, G., and Bussman, H. A phase-ability model of mathematical problem solving. *Journal for Research in Mathematics Education*, 1980, *11*, 179–189.

Lester, F. Research on mathematical problem solving. In R. Shumway (Ed.), *Research in mathematics education*. NCTM Professional Reference Series. *Reston, Virginia: NCTM, 1980.*

Levine, H., and Mann, K. The "negotiation" of classroom lessons and its relevance for teachers' decision-making. Paper presented at the annual meeting of the American Educational Research Association, Los Angeles, 1981.

McClellan, J. Philosophical redirections of educational research: Individualized instruction: A projection. National Society for the Study of Education Yearbook, 1972.

McConnell, J. Relationships between selected teacher behaviors and attitudes/achievements in algebra class. Paper presented at the annual meeting of the American Educational Research Association, New York, 1977.

Medley, D. The effectiveness of teachers. In P. Peterson and H. Walberg (Eds.), *Research on teaching: Concepts, findings, and implications*. Berkeley, California: McCutchan Publishing Corp., 1979.

Medley, D., and Mitzel, H. The scientific study of teacher behavior. In A. Bellack (Ed.), *Theory and Research in Teaching*. New York: Bureau of Publications, Teachers College, Columbia University, 1963.

Metz, M. *Classrooms and corridors: The crisis of authority in desegregated secondary schools*. Berkeley: University of California Press, 1978.

Miller, N. Personality differences between black and white children in the Riverside school study. Paper presented at the meeting of the Society of Experimental Social Psychology, Minneapolis, Minnesota, 1970.

Miller, N. Summary and conclusions. In H. Gerard and N. Miller (Eds.), *School desegregation: The long-range study*. New York: Plenum Press, 1975.

Mitzel, H., and Gross, C. The development of pupil growth criteria in studies of teacher effectiveness. *Educational Research Bulletin*, 1958, *37*, 178–187.

Moos, R. A typology of junior high and high school classrooms. *American Educational Research Journal*, 1978, *16*(1).

Murphy, P., and Brown, M. Conceptual systems and teaching styles. *American Educational Research Journal*, 1970, 7(4), 529–540.

Newsom, R., Eischens, R., and Looft, W. Intrinsic individual differences: A basis of enhancing instructional programs. *Journal of Educational Research*, 1972, 65, 387–392.

Peterson, P. Interactive effects of student anxiety, achievement orientation, and teacher behavior on student achievement and attitude. *Journal of Educational Psychology*, 1977, 69, 779–792.

Peterson, P., Janicki, T., and Swing, S. Ability by treatment effects on children's learning in large-group and small group approaches. *American Educational Research Journal*, 1981, 18(4), 453–473.

Peterson, P., and Wilkinson, L. Merging the process-product and social linguistic paradigms: Research on small-group processes. Paper presented at the annual meeting of the American Educational Research Association, New York City, 1982.

Polya, G. *How to solve it* (2nd Ed.). New York: Doubleday, 1957.

Power, C. *A multivariate model for studying person-environment interactions in the classroom*. Technical Report No. 99. Center for Research in Social Behavior, University of Missouri, Columbia, 1974.

Rabinowitz, W., and Rosenbaum, I. Failure in the prediction of pupil-teacher rapport. *Journal of Educational Psychology*, 1958, 49, 93–98.

Remmers, H. Rating Methods in Research on Teaching. In N. Gage (Ed.), *Handbook of Research on Teaching*. Chicago: Rand McNally, 1963.

Riedesel, A. Verbal problem solving: Suggestions for improving instruction. *Arithmetic Teacher*, 1964, 11, 312–316.

Rosenshine, B. *Teaching behaviors and student achievement.* London: The National Foundation for Educational Research, 1971.

Rosenshine, B. Content, time, and direction instruction. In P. Peterson and H. Walberg (Eds.), *Research on teaching: Concepts, findings and implications*. Berkeley, California: McCutchan Publishing Corp., 1979.

Rosenshine, B., and Furst, N. The use of direct observation to study teaching. In R. Travers (Ed.), *The second handbook of research on teaching*. Chicago: Rand McNally, 1973.

Rosenthal, R., and Jacobson, L. *Pygmalion in the classroom: Teacher expectation and pupils' intellectual development*. New York: Holt, Rinehart and Winston, 1968.

Schmeck, R., Ribich, F., and Ramanaiah, N. Development of a self-report inventory for assessing individual differences in learning processes. *Applied Psychological Measurement*, 1977, 1, 413–431.

Schoenfeld, A. Measures of problem-solving performance and of problem-solving instruction. *Journal for Research in Mathematics Education*, 1982, 13, 31–49.

Schwab, J. The practical: A language for curriculum. *School Review*, 1969, 1–23.

Shavelson, R., and Stern, P. Research on teacher's pedagogical thoughts, judgments, decisions, and behavior. *Review of Educational Research*, 1981, 51(4), 455–498.

Shipp, D., and Deer, G. H. The use of class time in arithmetic. *Arithmetic Teacher*, 1960, 7, 117–121.

Shuster, A., and Pigge, F. Retention efficiency of meaningful teaching. *Arithmetic Teacher*, 1965, *12*, 24–31.

Silberman, C. *Crisis in the classroom: The remaking of American education*. New York: Random House, 1970.

Slavin, R. Cooperative learning and the alterable elements of classroom organization. Paper presented at the annual meeting of the American Educational Research Association, New York City, 1982.

Soar, R. An integrative approach to classroom learning. Public Health Service Grant No. 5–R11 MH 01096 and National Institute of Mental Health Grant No. 7–R11 MH 02045. Temple University, Philadelphia, Pennsylvania, 1966.

Solomon, D., and Kendall, A. J. *Final report: individual characteristics and children's performance in varied educational settings*. Chicago: Spencer Foundation, 1976.

Sperry, L. *Learning performance and individual differences*. Glenview, Illinois; Scott, Foresman and Co, 1972.

Stallings, J. Allocated academic learning time revisited, or beyond time on task. *Educational Researcher*, 1980, *9*, 11–16.

Stephens, J. *The process of schooling*. New York: Holt, Rinehart and Winston, 1967.

Suydam, M., and Weaver, J. *Using research: A key to elementary school mathematics*. Center for Cooperative Research with Schools, Pennsylvania State University, 1970.

Thelen, H. *Classroom grouping for teachability*. New York: Wiley, 1968.

Thies, A. A Brain-behavior Analysis of Learning Styles. In James Keefe (Ed) *Student Learning Styles: Diagnosing and Prescribing Programs*. Reston, VA.: National Association of Secondary School Principals, 1979.

Torrance, E., and Parent, E. *Characteristics of Mathematics Teachers that Affect Students' Learning*. Cooperative Research Project No. 1020. Minnesota School Mathematics and Science Center, Institute of Technology, University of Minnesota, 1966.

Walberg, H. (Ed.). *Educational environments and effects*. Berkeley, California: McCutchan Publishing Corp., 1979.

Webb, N. Predicting learning from student interactions: Defining the variables. Paper presented at the annual meeting of the American Educational Research Association, New York City, 1982.

Weinstein, C. The physical environment of the school: A review of the research. *Review of Educational Research*, 1979, *49*, 4.

Whitzel, J., and Winne, B. Individual differences and mathematics achievement: An investigation of aptitude-treatment interactions in an evaluation of three instructional approaches. Paper presented at the annual meeting of the American Educational Research Association, San Francisco, 1976.

Wickelgren, W. *How to solve problems*. San Francisco: Freeman, 1974.

Withall, J., and Lewis, W. Social interaction in the Classroom. In N. Gage (Ed.), *Handbook of Research on Teaching*. Chicago: Rand McNally, 1963.

Witkin, H., Dyk, R., Faterson, H., Goodenough, D., and Kays, S. *Psychological differentiation*. New York: Wiley, 1962.

Zahn, K. G. Use of class time in eighth grade arithmetic. *Arithmetic Teacher*, 1966, *13*, 113–120.

Author Index